V&R unipress

Interfacing Science, Literature, and the Humanities / ACUME 2

Volume 6

Edited by
Vita Fortunati, Università di Bologna
Elena Agazzi, Università di Bergamo

Paola Spinozzi / Alessandro Zironi (eds.)

Origins as a Paradigm in the Sciences and in the Humanities

With 8 figures

V&R unipress

acume2

Education and Culture

Socrates

This book has been published with the support of the Socrates Erasmus program for Thematic Network Projects through grant 227942-CP-1-2006-1-IT-ERASMUS-TN2006-2371/001-001 SO2-23RETH.

This project has been funded with support from the European Commission. This publication reflects the views only of the author and the Commission cannot be held responsible for any use which might be made of the information contained therein.

Questo volume è stato pubblicato con il contributo del programma Socrates Erasmus per i progetti di reti tematiche – 227942-CP-1-2006-1-IT-ERASMUS-TN2006-2371/001-001 SO2-23RETH.
Il progetto è stato finanziato con il contributo della Commissione europea.
La presente pubblicazione riflette le idee del solo autore e la Commissione europea non può ritenersi responsabile per l'uso che potrebbe essere fatto delle informazioni contenute al suo interno.

FSC Mix Produktgruppe aus vorbildlich bewirtschafteten Wäldern, kontrollierten Herkünften und Recyclingholz oder -fasern
Zert.-Nr. GFA COC 1229 www.fsc.org © 1996 Forest Stewardship Council

„Dieses Hardcover wurde auf FSC-zertifiziertem Papier gedruckt. FSC (Forest Stewardship Council) ist eine nichtstaatliche, gemeinnützige Organisation, die sich für eine ökologische und sozialverantwortliche Nutzung der Wälder unserer Erde einsetzt."

Bibliographic information published by the Deutsche Nationalbibliothek

The Deutsche Nationalbibliothek lists this publication in the Deutsche Nationalbibliografie; detailed bibliographic data are available in the Internet at http://dnb.d-nb.de.

ISBN 978-3-89971-759-4

Cover image: Constantin Brancusi, *Beginning of the World*, c. 1920, marble, nickel, silver, and stone (76.2 x 50.8 x 50.8 cm), Dallas Museum of Art, Foundation for the Arts Collection, gift of Mr. and Mrs. James H. Clark. Photograph © 2010 Dallas Museum of Art.
Printed in Germany

Contents

List of Figures 9

Paola Spinozzi / Alessandro Zironi
Introduction 11

Part I: Origins in Biological and Cultural Processes

Abstracts 25

Suresh I. S. Rattan
Origins of Ageing Mechanisms and Prospects for Anti-ageing Interventions 29

Alessandro Achilli / Ugo A. Perego
Mitochondrial DNA: A Female Perspective in Recent Human Origin and Evolution 41

Paola Spinozzi
The Origin of Art: Towards Humanistic-Scientific Theories and Methodologies 59

Andrea Grignolio / Vita Fortunati
Memory as the Origin of Biological and Cultural Identity in the Work of Samuel Butler 81

Bibliography Part I 105

Part II: Originating Communication: Behaviours, Emotions, Languages

Abstracts 115

Augusto Vitale
The Evolutionary Origin of Human Behaviour: the Case of Altruism . . . 119

Marco Dondi / Sergio Agnoli / Laura Franchin
A New Look at the Very Early Origins of Smiling 131

Laura Bafile
The Debate on the Origins of the Language Faculty in the Perspective of the Biolinguistic Programme 147

Keith Hoskin
Origins or Geneses? The Roles of Invention 163

Bibliography Part II 181

Part III: Quests for Origins, Origins of Narrations

Abstracts 191

Marcella Farioli
The Genesis of the Cosmos, the Search for the *Arche* and the Finding of *Aitia* in Classical Greek Culture 195

Alessandro Zironi
Searching for Origins: Indo-European Words and Nordic Representations 211

Marusca Francini
Origo Gentis and Cultural Memory. The Origins of England as an Emblematic Case Study 227

Marianne Børch
The Semantics of Originality as Indicator of the Cultural Paradigm Shift from Medieval to Modern 243

Jón Gunnar Þorsteinsson
Paradigm as an Origin: Or, Bearing Evidence to the Meaningful Logic of Cause and Effect 257

Bibliography Part III 269

Bio-bibliographical Profiles 275

Index 281

List of Figures

1. Schematic phylogeny of human mtDNA.
2. Human migration routes based on mtDNA data.
3. Le mur végétal, Musée du quai Branly, Paris, 2006. Architect: Jean Nouvel; GreenWall. Designer & Installation: Patrick Blanc.
4. Le mur végétal, Musée du quai Branly, detail.
5. Constantin Brancusi, *Beginning of the World*, c. 1920, marble, nickel, silver, and stone (76.2 x 50.8 x 50.8 cm), Dallas Museum of Art, Foundation for the Arts Collection, gift of Mr. and Mrs. James H. Clark. Photograph © 2010 Dallas Museum of Art.
6. Anish Kapoor, *Turning the World Inside Out*, 1995, cast stainless steel (3/3, 152.4 x 177.8 cm), Santa Barbara Museum of Art, Museum purchase, 20th Century Acquisition Fund and funds provided by Eli and Leatrice Luria and the Luria Foundation, Jon and Lillian B. Lovelace Jr., Smith Richardson and the Grace Jones Richardson Trust, and the SBMA Visionaries. Photograph © 2010 Santa Barbara Museum of Art.
7. José Bedia, *Empezando a echar raíces al otro lado (Beginning to grow roots on the other side)*, 2002, lithograph printed in brown, grey, black and white from four aluminium plates on cream Rives BFK paper (76.2 x 112.4 cm), Boston, Museum of Fine Arts, Lee M. Friedman Fund, 2003.625. Photograph © 2010 Museum of Fine Arts, Boston.
8. Samuel Butler's theory of unconscious memory.

Paola Spinozzi / Alessandro Zironi

Introduction

> The most beautiful thing we can experience is the mysterious. It is the source of all true art and science.
> Albert Einstein, *The World as I See It*, 1940

I. The Intuition and the Pursuit

The *raison d'être* of this volume can be enlightened by focusing on the intuition that the multi-layered-ness and ductility which characterises the concept of origins appeals both to scientists and humanities scholars. From an etymological point of view the noun 'origin' relates to '1. The act or fact of arising, springing, from something [...]; 2. That from which anything arises, springs or is derived'. [1] The adjective 'original' means 'pertaining to the origin, beginning, or earliest stage of something' [2] and bears connotations such as pristine, primeval, primary. Clearly enough, the indication that there is a source, a commencement, and the condition of being primitive, authentic, archetypal, archaic, essential, pure, allow for remarkable semantic mobility. The wide range of significations renders the concept of origins perfectly suited to host interdisciplinary dialogue.

The assumption that origins can be defined as a hermeneutic paradigm in the humanities and in the sciences is explored in relation to specific theoretical frameworks and research methodologies. By investigating how origins have been conceptualised in different domains of knowledge – biology, primatology, psychology, linguistics, history of science, critical theory, classical studies, philology, literary criticism, strategy and accounting – a double movement has been generated: towards the very core of each discipline and beyond disciplinary boundaries. 'Which are the most productive theories and methods each discipline has elaborated for investigating origins?' 'Can they become transdisciplinary?' 'Which synergic enquiries can be devised in order to expand and share knowledge?': such questions have been addressed in order to verify if origins function as a powerful connector between scientific and humanistic

1 *Oxford Dictionary*, 2nd edn, 1989, vol. X, p. 933.
2 Ibid., ivi.

territories. The prospect of envisaging a context in which a geneticist and a classicist discuss the origins of humankind may well sound provocative. Indeed, one of the major aims pursued in studying origins has been the constitution of an heterogeneous research environment suited to foster discussion about the theoretical orientations and methodological tools which various domains of knowledge have regarded as distinctive and pre-eminent.

According to the different lines along which the research has been conducted, it could be said that there are scientific disciplines such as biology, primatology and psychology which share a predisposition to study origins adopting an evolutionistic framework; there are humanistic disciplines such as classical studies and history of religions which discuss the speculative attitudes underlying representations of origins and seem to be only marginally affected by scientific issues; there are disciplines such as linguistics, anthropology, sociology and astrophysics which had to face constraints and even censorship when approaching the topic of origins. In spite of the numerous attempts to involve scholars from these disciplines, only linguistics is represented here. While regretting the absence of the other ones, one may argue that, owing to specific historical controversies, origins are still regarded as a heated area of research which requires further internal discussion and clarification, before transdisciplinary dialogue becomes a fundamental objective. However, because origins involve a quest, which is central in every scientific field, the quest for origins cannot be ignored. It can be excluded, for reasons inherent in the discipline, but not ignored. Thus, explaining why and how various disciplines have responded to the topic of origins involves delving into their histories and cultural ideologies. Insight into the motivations which led certain disciplines to endorse or reject research on origins can be gained by reading the series of lectures kept at Darwin College, Cambridge, in 1986. [3] Origins of the universe are explored by Martin J. Rees, of the solar system by David H. Hughes, of complexity by Ilya Prigogine, of humans by David Pilbeam, of social behaviour by John Maynard Smith, of society by Ernest Gellner, and of language by John Lyons. It is easy to notice that the sciences are represented by outstanding authors, but the humanities occupy a marginal place. The present volume intends to balance the presence of scientists by including contributions from scholars of ancient, early, modern and contemporary literatures.

3 See A. C. Fabian, ed., *Origins: The Darwin College Lectures* (Cambridge: Cambridge University Press, 1988).

II. The Incubation

Our questions about origins germinated in Iceland, [4] whose geology has been studied as a model for the natural sciences and whose history and literature have been identified as the core of Northern European cultural identity. Scientists have studied the peculiar genetic heritage of the Icelanders and analysed the earth, fauna, and flora, while novelists, historians and sociologists have praised Old Norse mythology, poetry and sagas, regarded as the substrate that has nurtured the sense of a shared Teutonic cultural heritage.

However, in our working hypothesis beginnings are not deemed more important than developments. Indeed, when origins are found, mutations are also found. In Iceland the unceasing outburst of primary elements, far from preserving the original geological features, has left deep marks and produced radical changes. Furthermore, the belief that Iceland guards a pure, unspoiled, untainted, and thus unquestionable Germanic past discloses in fact the germs for incessant re-enactings of memory and history through evolving ideologies. Iceland exemplifies how a model of origins necessarily raises issues about its evolution.

Journeys into origins, intended in the widest sense and thus encompassing not only enquiries into the genesis of the universe, the earth, humans, but also quests for the sources of culture, are motivated by a reappraisal of acknowledged values, norms and conventions, and by the necessity to speculate on alternative social and cultural models.

While evoking very remote or bygone epochs, origins mark a beginning. They presuppose a 'before' that, compared with 'afterwards', could be exalted to an ideal excellence, but could also be criticised as less developed and imperfect. Thinking about origins engenders a wavering between backwards and forwards and calls for a comparison between different stages or ages: beginnings may be better because they are archetypal, original, pure, unspoiled, or may be worse, because they are embryonic, incomplete, unrefined, unpolished, or even coarse.

In disclosing views about the origin of being, narrations of the beginning address ontological questions. Glowing with wonder, cosmologies from the ancient East and pre-Socratic speculations investigate the beginning of reality. Divinities are creators in all mythical representations produced by the most ancient societies such as the Sumerian, Babylonian, Assyrian, Egyptian. The *Book of Genesis*, Hesiod's *Theogony*, Plato's *Timeus* call for evidence of a commencement which is complex and folded in silence. Even Aristotle postulates a

4 Reykjavik was the venue of the conference on *The Cultural Reconstruction of Places*, promoted by the European Thematic Network Project *ACUME. Cultural Memory in European Countries* and held between 24 and 26 June 2005.

Prime Mover who, though immovable, starts the mechanism of being. The relationship between divinity and creativity is deep. Divinity must create; God is the creator by definition.

Enquiries into origins involve not only exploring mechanics of creation, but also addressing abysmal issues. The question of questions, as ancient as pre-Socratic or pre-Platonic thought, is why there is not nothingness. The ancient Greeks loathed nothingness and did not have words for an absolute *ex nihilo*. The concept of nothing thus acquired a figurative value and was often personified as Death, Sleep, senseless confusion. Hesiod does not explain how, but Chasm gave birth to Erebos and black Night. [5] Creation thus becomes a procreation, an act of fertility.

Islamic mysticism is rich in models of creation which form one of the most complex theosophies. Moreover, in Sufism feminine components of creativity are very prominent: the *Sophia aeterna* revealed to Ibn Arabi as the source of poetic inspiration and divine initiator of love takes the shape of a young woman. The possibility of a *creatio ex nihilo* is not contemplated: the living beings are the expression of a infinite range of possible beings in which the Divine Being reveals his emanation.

The Jewish Masters maintain that this world was not the very first creation of God, who had generated, and then destroyed, other worlds. He did so because, after having created them with justice, he soon realized that justice was not enough to guard them. He then saved this world because, tired of destruction, he created it with mercy, hoping that it would remain. Thus, the first word of the Bible is 'bereschit', in the beginning. It does not indicate creation from nothingness, as part of Christian theology maintained, but rather the beginning of the relationship between God and world. Before the creation there were emptiness, darkness and the presence of God. With his word ('and God said') he contrasts nothingness and separates light from darkness, wind from water, the time of the day from the time of the night. In one week, the duration of time, God accomplishes creation.

In a midrash by the Palestinian Master Tanchumah bar Abba (IV century A.C.) it is told that the primordial and eternal Torah, by which the Eternal created and is eternally creating the universe, is written by black fire on the white fire of his divine lap. The letters of Torah are thus made of fire, and as the flame is constantly changing its form, so the universe changes, even though it is one and unique for all creatures and its nature is eternally immutable. Notwithstanding its immutability, the universe changes, according to the different perspectives and moods of the single creatures who contemplate it, participating in its

5 See Hesiod, *Theogony, Works and Days, Testimonia*, ed. and trans. by Glenn W. Most (Cambridge – London: Harvard University Press, 2006), ll. 116 – 120; 123 – 133, p. 13.

existence. Every view, deriving from single and always changeable contemplations, has to be saved in individual memory because, owing to its uniqueness, it is a precious wealth.

The concept of nothingness has permeated speculative thought in modern and contemporary times. It was Pascal who in his *Pensées* (1670) introduced the issue of nothingness:

He who thus considers man will be affrighted at himself, and regard himself as if suspended in his natural body between these two abysses of infinitude and nothing; he will tremble to behold these marvels [...].
For after all, what is man in nature? A cipher, compared to infinity; a whole, compared to a cipher; a medium between nothing and all. Infinitely distant from comprehending extremes, equally incapable of seeing the nothing from which he is drawn or the infinite in which he is engulfed, the end of things and their principles are for him invincibly hidden in impenetrable secrecy. [6]

Pascal's reflections are pervaded by a sense of yearning, awe, and anxiety: attempts at understanding the origin and the end are doomed to fail, because they are inscribed in God's design, which cannot be encompassed by human faculties. Simply, human beings cannot know, because they are unable to grasp the breadth of both infinity and nothingness. Their extension is incommensurable.

In *Wissenschaft der Logik* (*The Science of Logic*, 1812–1816) Hegel emphasized that our ability to use nothingness as a predicate, which generates the apparent oxymoron of the preposition 'es gibt nichts', is essential to any serious epistemology. According to Hegel, the beginning is not pure nothingness, but pure non being, from which being will be born. [7]

The beginning is not pure nothing, but a nothing from which something is to proceed; therefore being, too, is already contained in the beginning. The beginning, therefore, contains both, being and nothing, is the unity of being and nothing; or is non-being which is at the same time being, and being which is at the same time non-being. [8]

6 *Pascal's Pensées, or Thoughts on Religion*, trans. and ed. by Gertrude Burford Rawlings (Mount Vernon, N.Y.: Peter Pauper Press, 1946), p. 105.
7 Georg Wilhelm Friedrich Hegel, *Science of Logic* (London: Routledge, 2002). See *Book One: The Doctrine of Being*, in particular *With What Must the Science Begin*, pp. 67–78; *Chapter One. Being*, pp. 82–108, and *Book Two: The Doctrine of Essence*, in particular *Chapter I. Illusory Being*, pp. 394–408.
8 Georg Wilhelm Friedrich Hegel, *With What Must the Science Begin*, p. 73.

In the beginning there is relation, that is possibility. In *Was ist Metaphysik?* (*What is Metaphysics?*, 1929), Heidegger maintained that Hegel's proposition about pure Being and pure Nothing being one and the same is correct:

Being and Nothing hang together, but not because the two things from the point of view of the Hegelian concept of thought are one in their indefiniteness and immediateness, but because Being itself, is finite in essence and is only revealed in the Transcendence of Da-sein as projected into Nothing. [9]

Heidegger disrupted ordinary language and syntax barriers by turning 'nichts' into the verb 'nichten'. Though similar to 'vernichten', 'to annihilate', it acquires a different meaning. As M. J. Inwood explains:

Nichtung, 'noth-ing, nihilating', is what the Nothing does, just as *die Welt weltet*, 'the world worlds' or light light(en)s. The nothing is used positively: Heidegger is not saying 'There is not anything that noths', but 'Something noths, namely the Nothing'. The Nothing is not to be explicated in terms of negation: 'it is the origin of denial'. [10]

The Nothing does something: it negates. It is thus active and effective, insofar as it generates a negation. A thorough perusal of Hegel's and Heidegger's conceptualisations of being and nothingness allowed Sartre to expand on their interrelatedness in *L'être et le néant. Essai d'ontologie phénoménologique* (1943). [11] Being and nothingness become absolutely inseparable: in order to be envisioned, nothingness must be related to being, and as soon as that happens, it disappears as such and leads us back to being. Nothingness can be nihilated only if being is presupposed. Nothingness is neither before nor after nor outside being, but dwells in it. [12]

[...] no question could be asked, in particular not that of being, if negation did not exist. But this negation itself when inspected more closely referred to us back to Nothingness as its origin and foundation. In order for negation to exist in the world

9 Martin Heidegger, *What is Metaphysics*, trans. by R. F. C. Hull and Alan Crick, in *Existence and Being by Martin Heidegger*, with an Introduction by Werner Brock (Chicago: Henry Regnery Company, 1949), pp. 353–99.

10 Michael Inwood, 'Nothing and negation', in *A Heidegger Dictionary* (Malden: Blackwell, 1999), p. 145.

11 Jean-Paul Sartre, *Being and Nothingness. An Essay on Phenomenological Ontology*, trans. by Hazel E. Barnes, Introduction by Mary Warnock (London, Routledge: 2000), *Part One. The Problem of Nothingness*, pp. 3–70, in particular *Chapter One. The Origin of Negation*, pp. 3–45.

12 Jean-Paul Sartre, *III. The Phenomenological Concept of Nothingness*, in *Chapter One. The Origin of Negation*, pp. 16–21 (p. 21).

and in order that we may consequently raise questions concerning Being, it is necessary that in some way Nothingness be given. We perceived then that Nothingness can be conceived neither *outside* of being, nor as a complementary, abstract notion, nor as an infinite milieu where being is suspended. Nothingness must be given at the heart of Being, in order for us to be able to apprehend that particular type of realities which we have called *négatités*. But this intra-mundane Nothingness cannot be produced by Being-in-itself; the notion of Being as full positivity does not contain Nothingness as one of its structures. We can not even say that Being excludes it. Being lacks all relation with it. Hence the question which is put to us now with a particular urgency: [...] *where does Nothingness come from?* [13]

It is a question that in this volume has been rephrased as *what does Nothingness bring forth?* by pursuing the idea that nothing is what leads to being which keeps memory of nothing. Nothing-cum-being is a synonym for origin based on the assumption that a dialogue among scientific and humanistic disciplines requires defining what each of them is, but also envisaging a wider perspective of not being, being and becoming.

III. The Routes

The reflections about the idea of origins triggered by the journey to Iceland coalesced in a research project presented at the Villa Vigoni, on the Lake Como, in November 2006, during the first brainstorming meeting of the European Thematic Network devoted to *Interfacing Sciences, Literature and Humanities.* Soon afterwards, at the first General Meeting, which took place in Bertinoro (Italy) in March 2007, sub-project 3 identified origins as a research topic which, owing to its malleability, could stimulate responses from both scientists and humanists.

The project was further discussed at a preparatory meeting held at the University of Bologna in June 2007: keywords connected with the idea of origin were identified and research areas to be involved were defined. It is rewarding to notice that most of the topics envisaged at the time have actually been tackled: *Homo sapiens* features in two chapters, one by Alessandro Achilli and Ugo A. Perego, the other by Paola Spinozzi; the relationships between human and animal behaviour have been investigated by Augusto Vitale; language is delved into by Laura Bafile; an assessment of what is biological and what is cultural in memory is offered by Vita Fortunati and Andrea Grignolio; the emotional sphere is analysed by Marco Dondi, Sergio Agnoli and Laura Franchin; how

13 Jean-Paul Sartre, *III. The Phenomenological Concept of Nothingness*, in *Chapter One. The Origin of Negation*, pp. 21–22.

stocasticity and environment affect human lifespan is examined by Suresh I. Rattan; Marcella Farioli, Alessandro Zironi, Marusca Francini and Marianne Børch explore how narrations and myths of foundation came into existence and were transmitted; Keith Hoskin discusses how modern forms of disciplining individuals and populations were designed and applied.

Scientists and humanists first met at a workshop entitled *Decoding the Origins*, held at the University of Warwick in December 2007: twelve representatives of specific disciplines, many of whom present in this volume, discussed how origins have been explored in different branches of scholarship, which frameworks have been devised, why certain issues have been addressed or disregarded. After having investigated a wide range of conceptualisations, the conference held in Ferrara in May 2008 was then devoted to identifying common areas of investigation, points of divergence and travelling concepts. Bold as it may seem, our belief is that the task we set upon ourselves has been fulfilled. This volume will prove if our self-assessment may be considered fair. Certainly the authors bear evidence to the initial assumption that the concept of origins fosters the identification of a shared trans-disciplinary terrain where scientists and humanists can meet and experiment on integrated modes of enquiry.

IV. The Book

Hopefully the statement that *Origins as a Paradigm in the Sciences and in the Humanities* is not a collection of proceedings will sound genuine, if it is qualified not only by the indication that sections were reconsidered and further collaborations requested, but also by the specification that the research about origins which takes shape in this volume has evolved through the last five years.

Part I, entitled *Origins in Biological and Cultural Processes*, begins by exploring the origins of populations and their expansion through continents and times. Population geneticists have reconnected the whole mankind to a primeval haplogroup from Africa: far from being a discovery which pertains exclusively to the sciences, it clearly raises issues which have an enormous cultural import, because it undermines deeply rooted ideas about racial purity and calls for a wider conceptualization of ethnic groups. The evolutionary approach adopted by population genetics is given further breadth through theories of the origin and evolution of human distinctive traits such as ageing, memory, artistic perception and expression. Nineteenth-century theories of memory offer paradigmatic instances of enquiries into the core of the human being: the belief that experiences and memories of ancestors and parents are encapsulated in our cells, deeply undermined by Darwinian theories, reveals the necessity to detect not only origins, but also permanence through transmission and inheritance.

The chapters included in Part I show how primordial, unchangeable constituents of the human species affect areas of behaviour traditionally considered as individual and arbitrary. What emerges is a complex bio-cultural map of the world, in which life is defined through the constant interplay between hereditary, inborn inclinations and skills gradually acquired. Indeed, if scientists and humanists join their efforts, the interactions between genetic predispositions and the natural and cultural milieu, which can be regarded as a major research area in numerous disciplines, will be studied in its complex multi-faceted-ness.

Part II, devoted to *Origins of Communication: Behaviours, Emotions, and Languages*, focuses on specific human modes of interaction such as empathy and altruism, smiling, language and management. Behavioural patterns and emotional responses prove to be universal to a considerable extent. Again, what requires deeper scrutiny are the processes by which certain innate human attitudes and gestures cease to be essentially automatic and instinctive and begin to be voluntary, because their functionality and social utility have been apprehended. Language can be explained precisely as a human faculty being stimulated by a special intertwinement of inborn and trained abilities. After having examined the anatomical structure of the human phonetic apparatus, linguists have reached the conclusion that the ability to speak is due to a gradual cultural process activated within a prearranged anatomical structure. Not only a basic form of expression as language, but also mediated forms of communication as marketing and strategic management testify to the biological/environmental nexus. Emotional components interfere with transmission and reception of executive tasks: from the propulsive centre, where decisions originate, to the periphery, where they should be enacted, information is inevitably affected by the emotions of all those who are involved in the process.

The origin of altruistic behaviours and linguistic skills in human and non-human primates, of smiling in early infants, of accounting as a form of social control bring further evidence supporting the theory that the study of the interaction between innate dispositions and developmental and cultural transformations is an endeavour requiring synergic contributions from the sciences and the humanities.

Part III, entitled *Quests for Origins: Mappings, Narrations and Representations*, explores how the genesis and development of the universe, earth and man were narrated in literary sources dating to different historical and cultural periods. A deeper insight into narrations of origins in ancient epochs can be gained by adopting a comparative approach: cosmogonies, theogonies and anthropogonies in classical and Old Norse culture provide emblematic case studies.

History summoned to show origins can be powerfully manipulated to fulfil different purposes, as appears from different chronicles of the same events. Constructed in order to convey specific ideological contents, historical accounts of the formation of a country intertwine facts and mythical elements. The origins of Anglo-Saxon England as related in different texts are a remarkable case study. The definition of originality also acquires different meanings in different historical periods. In the Middle Ages invention entailed the certainty that everything already existed and nothing could be newly created, because creation had already been accomplished. The beginning of the modern age entailed overcoming the medieval idealisation of perfect immutability. Originality in relation to invention thus allows for detecting semantic changes which enclose changing world views and, consequently, paradigms shifts.

The final chapter of the volume delves into the concept of paradigm. Does it work, applied to origins? If we consider that a paradigm is a clearly defined and coherent set of concepts, we will have to admit that the definition only partially applies to the concept of origins. Owing to its flexibility, it could be defined as a hyper-paradigm, because it is connected with a cluster of paradigms formed within different disciplines. Indeed, its ubiquitous, flexible nature has been fundamental for establishing a sustained dialogue between the sciences and the humanities.

Achieving cohesion while pointing out specificities can be regarded as a goal from which interesting results have ensued. While the chapters well portray the state of the arts on the idea of origins articulated by various disciplines, they reveal that identifying trans-disciplinary paths has not always proved an easy task. Making interface is of course an intriguing challenge, but the questions remains as to who, between humanities scholars and scientists, can bring 'the other' in more easily and thought-provokingly.

V. Coalescences

The *horror vacui* felt by ancient civilisations when facing questions about the creation of cosmos has resurfaced in the sense of suspension with which contemporary age responds to relativism and to the awareness that no new theorization can elude the notion of chance. The belief in immutability was shattered in the nineteenth century: what for thousands of years had been considered unmodifiable – universe, nature, man, life – was instead subject to chance. Nowadays the structure of the DNA and the mechanisms of life, ageing and death would not be understandable without the notion of stochasticity. The acknowledgement that only the concept of complexity can encompass the acquisition of knowledge and the still partial understanding about the human body, the laws of nature, and universe, are dizzying.

Mapping the world with clear indications about origins can be considered as still another re-enactment of the search for the primeval status, a striving for the correspondence between microcosm and macrocosm. Origins are still alluring. Because they generate endless speculations and do not cease to prompt quests, we would almost be tempted to say that the abyss from which all was created is, above all, an invention conjured up by the human mind. However, explanations which do not dwell on the ineffable, bewildering components of human nature prove more productive. Origins are a crucial theme in a world which is remodelling itself while coping with globalization. Western culture, after having prevailed for so many centuries, has acknowledged that its role must be revised and reframed. The last G8's meeting held in Italy in July 2009 emphasised that inter-governmental institutions should be re-constituted in order to include non-western countries such as China and India. In this context of transformation, a sustained debate on the idea of origins can help envision a new world.

Interestingly, the search for origins seems to subvert a notion associated with the Two Cultures, according to which science should pursue the progress of mankind, while humanistic studies, framed within a historical perspective, are supposed to focus on the past. In fact, the geneticists studying the DNA, the gerontologists researching ageing, the primatologists and psychologists experimenting on innate or acquired components of behaviours offer scientific views of origins which explore the biological history of humans in order to understand how they 'work' now.

It would be absurd to define what being humans means by attempting to grasp the degree zero, to fabricate a state totally unaffected by external factors: an environment and the ability to shape it are fundamental for human life. New disciplines originate from continuity of knowledge, new research orientations develop with the aid of methodological instruments shaped in specific fields and re-shaped in others. Thus, a productive endeavour to be shared by the sciences and the humanities is the assessment of what the outset – any possible outset – was, how it has metamorphosed while being transmitted to the present and how it will continue to change in the future.

Part I: Origins in Biological and Cultural Processes

Abstracts

Suresh I. S. Rattan

Origins of Ageing Mechanisms and Prospects for Anti-ageing Interventions

Healthy ageing and longevity depend on successful and dynamic interactions among biological, psychological and environmental factors. The origins of mechanisms of ageing do not lie in the genetically controlled evolutionary processes involving any real gerontogenes. Instead, it is the failure of maintenance and repair systems constituting the 'homeodynamic space' that underlies the biological basis of ageing, age-related diseases, and eventual death. This view facilitates setting up a framework for understanding, researching and developing effective and realistic strategies for ageing intervention, prevention and therapies. Gene therapy, stem cells, and modulation through functional foods, nutriceuticals, cosmeceuticals and life style alterations are examples of such strategies. A promising novel anti-ageing approach is to strengthen the homeodynamic ability of self-maintenance, by using mild stress-induced hormesis through physical, nutritional and mental hormetins.

Alessandro Achilli / Ugo A. Perego

Mitochondrial DNA: A Female Perspective in Recent Human Origin and Evolution

With its 16,569 base pairs, the mitochondrial genome is very short when compared to the nuclear genome. Nevertheless, the study of its variation has contributed tremendously to our understanding of the origin of modern humans and the routes they followed as they colonized the world. Two peculiar characteristics make the mitochondrial DNA (mtDNA) a valuable tool in population studies: firstly, contrary to traditional genetic markers found in the nuclear DNA, mtDNA does not recombine, but it is inherited exclusively along the unbroken maternal lineage; secondly, because its mutation rate is higher than nuclear genes, its process of molecular differentiation was relatively fast and occurred mainly during and after the process of human colonization and diffusion into various regions and continents. Therefore, the different subsets of mtDNA variation (called haplogroups) tend to be restricted to different geographic areas and population groups.

The reader will be guided through the major events that characterized the history of modern humans over the last 200,000 years from the female (mitochondrial) perspective. In fact, thanks to the study and analysis of mtDNA variation in modern and ancient DNA samples, scientists have been able to derive important clues about the African origins of *Homo sapiens* (the famous 'mitochondrial Eve'), the first spread from the African continent (known as the 'Out of Africa' theory), and the subsequent migration routes followed by our ancient ancestors toward the colonization of the whole world.

For lack of space, we will examine three stages in order to exemplify the process of modern humans' expansion. First we will focus on the primordial 'step' from the African continent, then we will follow the migration into Europe, where our first ancestors came face to face with their distant cousins, the Neanderthals, and finally, jumping from the Paleolithic to the Neolithic via the Last Glacial Maximum, we will reach North-Central Italy, where recent mtDNA data revealed the history and legacy of the great Etruscan civilization. These selected examples will demonstrate how scientific analysis of mtDNA can foster the development of new interfaced methods for researching the origin and diaspora of modern humans.

Paola Spinozzi

The Origin of Art: Towards Humanistic-Scientific Theories and Methodologies

The origins of Art expose the paradoxes of Western critical paradigms by showing how beauty, constantly associated with purity, truth and integrity, is a concept encrusted with the values as well as the biases which have marked the history of Western civilization.

The characteristics we attach to Western 'popular' art exhibit remarkable similarities with the features we note in the art of 'non-Western', 'indigenous' cultures. 'Popular' is the label conventionally assigned to woodcarving, embroidery, painted glass, metal working and every other form of craft, while 'non-

Western' and 'indigenous' are definitions utilized, not less conventionally, to encompass, and classify, the cultures of Africa, Asia, Oceania and Americas and to include artistic expressions not properly classifiable as 'pure' art. In fact, those artefacts may be made to be beautiful, but are certainly, and possibly primarily, intended for specific uses.

Trans-disciplinary enquiries into the origins of art can thrive on methods which, while maintaining the efficacy of historically and philosophically oriented critical modes, encompass archaeology, ethnology, cognitive sciences and biology. The concept of interface presupposes the activation of osmotic processes which can neutralise biased, hierarchical preconceptions about fine and lesser art, pure and applied creativity, us and them.

Because origins have been eluded by humanities scholars, the construction of a new cartography of the arts is necessary. The act of mapping involves studying the interactions between innate biological predispositions which trigger responses to emotional stimuli and the cultural environment which enhances or thwarts them. The artistic process presupposes the expression of emotional drives, is never learned and repeated mechanically, entails the deployment of technical skills and creativity, and requires an adequate environment. Interfaced methodologies are worth developing because they can offer a substantial contribution towards an understanding of art as an intrinsic, and shared, rather than a privileged, almost exclusive, human expression.

Andrea Grignolio / Vita Fortunati

Memory as the Origin of Biological and Cultural Identity in the Work of Samuel Butler

Samuel Butler's (1835 – 1902) theory of organic memory can be regarded as an original contribution to the Victorian debate on the cultural/biological origins of man and on the development of individual personality. Generally considered as mutually independent, the themes of origins and self were regarded by Butler as two related phases of an uninterrupted process of development: the evolutionary and cultural history of mankind. His theory of organic memory considers instincts, and especially unconscious recollections, as a bridge between individuals and the experiences of their ancestors. Indeed, traits of personality shared by parents and offspring show that biological and cultural origins are transmitted.

Butler's humanistic and scientific knowledge will also be taken into account. A man of letters known for his dystopian fiction *Erewhon*, he decided to shift his interests towards biological topics after having been deeply impressed by *On the Origin of Species*. The book did not only prompt a reorientation of his work, but also played an important role in his life. Butler first became a great admirer and friend of Darwin, then a strong opponent of his materialistic and chance-driven explanation of nature, to which he responded with his own teleological theory of evolution. Butler's contribution to biological thought will also be assessed by examining his twenty-year effort to integrate his theory of organic memory with other anti-selective theories, such as the ones set forth by Georges-Louis Buffon, Jean-Baptiste Lamarck, Erasmus Darwin, and Ewald Hering.

Suresh I. S. Rattan

Origins of Ageing Mechanisms and Prospects for Anti-ageing Interventions

I. A Question of Life and Death

Sooner or later, all individuals die out. Although the apparent immortality of a population may overshadow the mortality of its individual members, the life-span of an individual is not unlimited. Lifespan of organisms ranges from a few hours to hundreds of years, and a vast variety of life histories unfold before the ultimate cessation of life. In biological systems, three main categories of degeneration – rapid, negligible and gradual [1] – can explain most types of life histories that lead to the death of an individual. For example, rapid degeneration is common among species that die during or soon after reproduction. Such organisms generally undergo a single bout of reproduction during their life and are called semelparous, for example salmon fish. At the other end of the scale are organisms in which the rate of degeneration is imperceptible over very long periods and can be considered as undergoing negligible senescence. Various vascular plants, such as bristlecone pine and other conifers; invertebrates, such as lobsters; and vertebrates, such as rockfish and tortoise, belong to this category. The third category, found most commonly in animals, involves the growth and development of the organisms to adulthood and to a period of reproduction followed by gradual and progressive ageing and senescence culminating in death. Generally, species with repetitively reproducing (iteroparous) life histories experience ageing after completing a period of reproductive fitness. It is in this category of organisms, which includes human beings, that the phenomenon of progressive and intrinsic ageing is best manifested during the limited lifespan of the organism.

The main questions are, therefore: (i) what are the mechanisms of ageing and death? (ii) How did these mechanisms originate during evolution? (iii) Are these processes accessible to interventions aimed to delay ageing and thus to increase

1 Caleb E. Finch, *Longevity, Senescence, and the Genome* (Chicago: The University of Chicago Press, 1990).

the lifespan indefinitely? In order to answer these questions, it is important to understand certain basic principles of life processes, which also provide clues towards understanding what prevents these processes from functioning for ever.

II. Life and Homeodynamics

All living systems, in contrast to the non-living systems, have the intrinsic ability to respond, to counteract and to adapt to external and internal sources of disturbance. The traditional conceptual model to describe this property is *homeostasis*, which has dominated biology, physiology and medicine since 1930. However, tremendous advances made in our understanding of the processes of biological growth, development, maturation, and reproduction have led to the realization that homeostasis model as an explanation is seriously incomplete. The main reason for the incompleteness of the homeostasis model is its defining principle of 'stability through constancy', which does not take into account new avenues of thought opened up by cybernetics, control theory, catastrophe theory, chaos theory, information and interaction networks, which constitute the modern biology of complexity.

Since 1994, the term *homeodynamics*, introduced by F. E. Yates, has been increasingly used – though it has not yet fully succeeded in replacing homeostasis. [2] The concept of homeodynamics accounts for the fact that the internal milieu of complex biological systems is not permanently fixed, is not at equilibrium, and is a dynamic regulation and interaction among various levels of organization. Almost in parallel with the development of the concept of homeodynamics, another term, *allostasis*, coined and introduced by P. Sterling and J. Eyer in 1988, [3] has also gained recognition and use. According to the allostasis model, 'stability through change' is the most realistic situation for living biological systems. The model also takes into account characteristics such as reciprocal trade-offs between various cells, tissues and organs, accommodative sensing and prediction with respect to the severity of a potential stressor, and the final cost of making a response and readjustment to bring about the necessary change. Every act of allostasis adds to the *allostatic load* in terms of, for example, unrepaired molecular damage, reduced energy deposits and progressively less efficient or less stable structural and functional components. Ageing, senescence

2 Francis Eugene Yates, 'Order and Complexity in Dynamical Systems: Homeodynamics as a Generalized Mechanics for Biology', *Math. Comput. Model.*, 19 (1994), 49 – 74.
3 *Allostasis, Homeostasis, and the Costs of Adaptation*, ed. by Jay Schulkin (Cambridge: Cambridge University Press, 2004).

and death are the final manifestations of imperfect and unsuccessful homeodynamics.

The key biochemical and physiological pathways and processes of maintenance and repair that constitute the homeodynamic machinery of a living system are:

1) The multiple pathways of nuclear and mitochondrial DNA repair, including those for maintaining the accuracy of the information transfer from DNA to RNA to proteins, and those for the removal of spontaneous lesions in DNA.
2) The processes for sensing and responding to intra- and extra-cellular stressors, such as heat shock response, stress hormones, and ionic fluxes.
3) The pathways for protein repair, such as the renaturation of proteins by chaperones, and the enzymic reversal of the oxidation of amino acids.
4) The pathways for the removal and turnover of defective proteins by proteasomes and lysosomes.
5) The antioxidative and enzymic defences against reactive oxygen species.
6) The processes for the detoxification of harmful chemicals in the diet.
7) The cellular and humoral immune responses against pathogens and parasites, including massive apoptosis (programmed cell death) after the completion of the cellular immune response.
8) The processes of wound healing, blood clotting and tissue/organ regeneration including angiogenesis (blood vessel formation).

In addition to the main categories of pathways and processes constituting the homeodynamic machinery, some other physiological processes can also be included as components of homeodynamics. These processes include temperature control, the epigenetic stability of differentiated cells, and fat storage and energy utilization. Together all these processes give rise to a 'homeodynamic space' within which an individual operates and survives, and beyond which its chances of succumbing to diseases and death increase. At least theoretically, these homeodynamic processes, which are also referred to as the longevity-assurance processes, can work efficiently for ever, and an organism can be immortal – which obviously is not the case.

III. Homeodynamics and Longevity

The so-called longevity assurance processes assure longevity of an organism only in accordance with its evolutionary life history and for the duration required to fulfill its Darwinian purpose of life – reproduction and continuation of generations. In the context of the origin of mechanisms of ageing, it is incorrect to assume that ageing and limited lifespan of an individual had some purpose or adaptive significance in terms of being advantageous for the species. In natural wild populations the probability of death by accidental causes, including disease and predation, is so high that there is never a significant number of long-lived individuals left that might require special mechanisms to terminate life for the sake of newly born individuals. Even if there were any life-terminating mechanisms that operated after a long period of survival, these would not be capable of resisting the spontaneous origin and evolution of non-ageing and immortal 'mutants', which in a given population would soon take over.

A more widely accepted view is that ageing occurs because natural selection is insufficient to prevent it, owing to its post-reproductive nature, or that senescence is a by-product of the expression of genes with early beneficial traits but deleterious and pleiotropic effects at later stages. Two major schools of thought (whose ideas are not mutually exclusive) in the non-adaptive theories of the evolution of ageing and lifespan are represented by *antagonistic pleiotropy*, [4] and the *disposable soma theory* based on Weismann's distinction between the soma and the germ line. [5] According to these theories, evolutionary forces have optimised conditions for efficient and successful reproduction either by (i) selecting for 'good' early genes that later have 'bad' effects, or (ii) selecting for efficient maintenance and repair of the germ cells at the cost of somatic maintenance. The disposable soma theory [6] considers the evolved life history and the structural and functional design of the organism to be the basis of the evolution of ageing.

Thus, evolutionary forces of natural selection have resulted in evolving mechanisms of maintenance that operate in concert with the complete structural (anatomical) and functional (physiological) design of the organism and assure the survival of the body until reproduction. This has been termed 'essential lifespan, ELS' [7] or 'warranty period'. [8] ELS can be considered as the 'natural lifespan of a species', and is distinct (and usually several-fold shorter) from the average lifespan for a cohort, and from the maximum lifespan observed for a single member of a species. For example, for humans, ELS is considered to be about 40 years; whereas at present the average lifespan in rich countries is between 80 and 85 years, and the maximum human lifespan recorded so far is 122 years, 5 months and 14 days!

---

4 Michael R. Rose, *Evolutionary Biology of Aging* (New York: Oxford University Press, 1991).

5 Thomas B. L. Kirkwood and Thomas Cremer, 'Cytogerontology since 1881: a Reappraisal of August Weismann and a Review of Modern Progress', *Hum. Genet.*, 60 (1982), 101–21.

6 Thomas B. L. Kirkwood, *Time of Our Lives* (London: Weidenfeld & Nicolson, 1999).

7 Suresh I. S. Rattan, 'Biogerontology: the Next Step', *Ann. N.Y. Acad. Sci.*, 908 (2000), 282–90.

8 Bruce A. Carnes, S. Jay Olshansky, and Douglas Grahn, 'Biological Evidence for Limits to the Duration of Life', *Biogerontology*, 4 (2003), 31–45.

IV. Biological Principles of Ageing: Present and Future Prospects

Biogerontology, the study of the biological basis of ageing, has so far gathered a large body of descriptive data about the age-related changes in organisms, organs, tissues, cells and macromolecules. These data clearly show that the progression and rate of ageing are highly variable in different species, in organisms within a species, in organs and tissues within an organism, in cell types within a tissue, in sub-cellular compartments within a cell type, and in macromolecules within a cell. Although these data imply that there are no universally applicable markers of ageing, three general principles of biological ageing can be derived. [9]

First, the evolutionary *life history principle* describes ageing as an emergent phenomenon taking place primarily in protected environments, which enable survival beyond the natural lifespan in the wild (discussed above). Therefore, from an evolutionary point of view, ELS is the canvas against which genetic selection and functional optimization unfold.

Second, the *non-genetic principle of ageing* specifies that ageing and longevity of an individual are not programmed in specific gerontogenes. Unlike development, which is a highly programmed and well-coordinated genetic process in the evolutionary life history of an organism, there is no genetic programme which determines the exact duration of survival of an organism. However, as regards the role of genes in ageing, there are some misunderstandings that need to be clarified. For example, a lack of specific gerontogenes according to the 'non-genetic principle' does not imply that no genes have any influence on survival, longevity and ageing. There is ample evidence from studies performed on yeast, fungi, nematodes, insects, rodents and humans that mutations in certain genes can either prolong or shorten the lifespan in various animals, and cause premature ageing syndromes in humans. [10] However, it is most important to realise that these genes did not specifically evolve to cause age-related accumulation of damage or to kill the organism. Since their involvement in influencing ageing and longevity is only indirect, they have been termed 'virtual gerontogenes', [11] or 'longevity genes'. [12] Studies on establishing an association

9 Suresh I. S. Rattan, 'The Science of Healthy Aging: Genes, Milieu, and Chance', *Ann N Y Acad Sci*, 1114 (2007), 1–10.

10 Huber Warner, 'Longevity Genes: from Primitive Organisms to Humans', *Mech. Age. Dev.*, 126 (2005), 235–42; George M. Martin, 'Genetic Modulation of Senescent Phenotypes in *Homo sapiens*', *Cell*, 120 (2005), 523–32; Maris Kuningas, Simon P. Mooijaart, Diana van Heemst, Bas J. Zwaan, P. Eline Slagboom, and Rudy G. J. Westendorp, 'Genes Encoding Longevity: from Model Organisms to Humans', *Aging Cell*, 7 (2008), 270–80.

11 Suresh I. S. Rattan, 'Gerontogenes: Real or Virtual?', *FASEB J.*, 9 (1995), 284–86; Id., 'Ageing, Gerontogenes, and Hormesis', *Ind. J. Exp. Biol*, 38 (2000), 1–5; Id., 'Theories of Biological Aging: Genes, Proteins and Free Radicals', *Free Rad. Res*, 40 (2006), 1230–38.

12 Huber Warner, 'Longevity Genes'.

between genes and longevity have reported that the genetic heritability of variance in lifespan is less than 35 %. [13] Several studies have reported an association between human longevity and single nucleotide polymorphisms (SNP) in a variety of genes in various biological pathways, including heat shock response, mitochondrial functions, immune response, cholesterol metabolism and others. [14] The complex processes of maintenance and repair involve hundreds of genes whose products and interactions give rise to a 'homeodynamic space', or 'buffering capacity', which is the ultimate determinant of an individual's chance and ability to survive and maintain a healthy state. [15]

The third principle of ageing derived from biogerontological studies is the *mechanistic principle*, which characterizes ageing as a progressive accumulation of molecular damage in nucleic acids, proteins and lipids. Since homeodynamic ability of a living system is primarily due to its maintenance and repair systems, the progressive failure of these processes is the biochemical basis of ageing and age-related diseases. [16] The future of biogerontological research lies in resolving the complex nature of gene-protein networks and the role of molecular damage and heterogeneity in the process of ageing.

V. From Understanding to Intervention: Anti-ageing or Healthy Ageing?

Once it is understood that ageing is an emergent phenotype due to the failure of homeodynamics, and is not due to the action of any harmful and death-causing mechanisms, our approach towards ageing interventions shifts from 'anti-ageing' to 'healthy ageing'. According to the three principles of ageing and longevity described above, ageing occurs in spite of the presence of complex pathways of maintenance, repair and defence, and that there is no 'enemy within' that causes it. All mechanisms and processes in the body have evolved to work towards its survival and maintenance of health, in accordance with the evolutionary history

13 Thomas E. Johnson, Kaare Christensen, and James W. Vaupel, 'The Quest for Genetic Determinants of Human Longevity: Challenges and Insights', *Nature Rev. Genet.*, 7 (2006), 436–48; Ripudaman Singh, Steen Kølvraa, and Suresh I. S. Rattan, 'Genetics of Longevity with Emphasis on the Relevance of HSP70 Genes', *Frontiers in Science*, 12 (2007), 4504–13.

14 'Genes Encoding Longevity'; 'Genetics of Longevity'; Beáta Bessenyei, Marietta Márka, Lázló Urbán, Margit Zeher, and Imre Semsei, 'Single Nucleotide Polymorphisms: Aging and Diseases', *Biogerontology*, 5 (2004), 291–300.

15 'Theories of Biological Aging'; Suresh I. S. Rattan, 'Increased Molecular Damage and Heterogeneity as the Basis of Aging', *Biol. Chem*, 389 (2008), 267–72.

16 'The Science of Healthy Aging'; 'Theories of Biological Aging'.

of the species. This viewpoint makes modulation of ageing different from the treatment of one or more specific diseases. [17]

As a biomedical issue, the biological process of ageing underlies all major human diseases, such as atherosclerosis, cancer, cardiovascular defects, cataract, diabetes, dementia, macular degeneration, neurodegeneration, osteoporosis and sarcopenia. Although the optimal treatment of every disease, irrespective of age, is a social and moral necessity, preventing the onset of age-related diseases by intervening in the basic process of ageing is the best solution for improving the quality of human life in old age.

In the case of a disease, such as a cancer of any specific kind, its therapy will, ideally, mean the removal and elimination of the cancer cells and restoration of the affected organ/tissue to its original disease-free state. If the same therapeutic approach is applied to 'treat' ageing, it raises a crucial question as to what the 'treatment' of ageing means, and to what original 'age-free' stage should one hope to be restored. Furthermore, the basic notion of 'disease' in clinical terms implies that it is only a proportion of the population that becomes inflicted with that condition. However, a condition such as ageing, which happens to all who survive longer than ELS, cannot be considered as a disease, and cannot be approached as such.

Another important implication of understanding ageing as the inefficiency and imperfections of homeodynamics is that the prospect of developing anti-ageing magic bullets must be abandoned. This also means abandoning enemy-oriented rhetoric, such as the 'war against ageing', 'defeating ageing', and 'conquering ageing'. Instead, interventions in ageing require a 'friend-oriented' approach and the use of a positive language such as maintaining health, achieving healthy ageing, successful ageing, and preserving the homeodynamics.

Ageing, diseases and eventual death are the consequences of imperfect homeodynamics. Ageing is characterized by a progressive shrinkage of the homeodynamic space, mainly due to stochastic occurrence and specific accumulation of molecular damage. [18] A shrinking homeodynamic space implies a declining buffering capacity and increased vulnerability, impaired functionality and increased chances of one or more diseases. While most anti-ageing approaches are targeted against the treatment of specific age-related diseases, and these are often more effective and immediate, they do not address the process of ageing itself. [19]

17 Suresh I. S. Rattan, 'Anti-ageing Strategies: Prevention or Therapy?', *EMBO Reports*, 6 (2005), S25-S29.

18 'Theories of Biological Aging'; 'Increased Molecular Damage and Heterogeneity as the Basis of Aging'.

19 Leonard Hayflick, 'Biological Aging Is No Longer an Unsolved Problem', *Ann. NY Acad. Sci*,

Therefore, rational anti-ageing strategies based on scientific evidence aim to slow down the ageing process by preventing and/or delaying physiological decline and regaining, to some extent, lost functional abilities. Such approaches may be either piecemeal (short term) or long term. For example, piecemeal interventions include organ replacement, stem cell injections, and external supplementations to regain youthful levels of hormones, enzymes and micronutrients. Although some of these therapies have demonstrated some clinical benefits in alleviating problems associated with severe deficiencies in old age, none of these piecemeal interventions really act on and modulate the ageing process itself. [20]

VI. A Formula for Eternal Youth?

Long term prevention or slowing down of ageing will require effective interventions to increase the homeodynamic space and to decrease the rate of its shrinkage due to accumulation of unrepaired molecular damage. Therefore, in order to modulate ageing and to achieve healthy old age and/or to achieve eternal youth, three main conditions need to be fulfilled, as represented by the formula $E = GMC^2$, discussed below. [21]

The first parameter *G* represents genes, which are the basis of survival, growth, development and ELS. If all the genes involved in creating the homeodynamic space, especially in the pathways of damage removal and repair, can be manipulated in such a way that their interactive accuracy and efficiency remains functional indefinitely, then the first requirement for eternal life is fulfilled. However, at present our knowledge about the numbers, variations, interactions and means of manipulating genetic networks is too meager to even conceptualize what may be possible to do in this respect.

The second parameter *M* represent milieu – the environment in which living systems operate and survive. The milieu in which genes and gene products function ranges from the intracellular molecular and ionic milieu to all other orders of organization including cellular, physiological, psychological, and societal. Almost all the ongoing work on ageing modulation and intervention at present is aimed at modifying the milieu either by replenishing those enzymes,

1100 (2007), 1–13; Id., 'Entropy Explains Aging, Genetic Determinism Explains Longevity, and Undefined Terminology Explains Misunderstanding Both', *PLoS Genet.*, 3 (2007), e220.

20 'Anti-ageing Strategies: Prevention or Therapy?'; 'Entropy Explains Aging, Genetic Determinism Explains Longevity, and Undefined Terminology Explains Misunderstanding Both'; S. J. Olshansky, Thomas T. Perls, 'New Developments in the Illegal Provision of Growth Hormone for 'Anti-aging' and Bodybuilding', *JAMA*, 299 (2008), 2792–94.

21 'The Science of Healthy Aging'.

hormones and other molecules such as antioxidants and micronutrients whose levels are reported to decrease during ageing, or by increasing the levels of such biochemicals through nutritional supplementation. Some of the main approaches include supplementation with hormones including growth hormone, dehydroepiandrosterone, melatonin and estrogen, and nutritional supplementation with synthetic and natural antioxidants in purified form or in extracts prepared from plant and animal sources. Claims for the benefits of intake of high doses of vitamins and various antioxidants and their supposed anti-ageing and life-prolonging effects have very little scientific evidence to back them. In contrast to this, nutritional modulation through caloric restriction (CR) has been shown to be an effective anti-ageing and longevity extending approach in rodents and monkeys. However, this is a highly debatable issue at present both in terms of the practicalities of defining CR and of applying CR in human beings in physiological and evolutionary contexts. [22]

Another promising area of milieu improvement for achieving healthy ageing and longevity is through stress-induced hormesis. [23] Hormesis in ageing is defined as the life supporting beneficial effects resulting from the cellular responses to single or multiple rounds of mild stress. [24] As discussed above, a critical component of the homeodynamic property of living systems is their capacity to respond to stress. In this context, the term stress is defined as a signal generated by any physical, chemical or biological factor (stressor), which in a living system initiates a series of events in order to counteract, adapt and survive. While successful and over-compensatory responses to low doses of stressors improve the overall homeodynamics of cells and organisms, incomplete or failed homeodynamic responses lead to damaging and harmful effects of stress, including death.

The paradigm for considering the applicability of hormesis in ageing intervention is the well documented beneficial effects of exercise, which at a biochemical level results in the production of potentially harmful substances such as free radicals, acids and aldehydes. [25] Mild stresses that have been reported to delay ageing and prolong longevity in various systems include temperature shock, irradiation, heavy metals, pro-oxidants, acetaldehyde, alcohols, hypergravity, exercise and food restriction. [26] Hormesis-like beneficial effects of

22 Eric Le Bourg and Suresh I. S. Rattan, 'Can Dietary Restriction Increase Longevity in All Species, Particularly in Human Beings? Introduction to a Debate Among Experts', *Biogerontology*, 7 (2006), 123–25.
23 Suresh I. S. Rattan, 'Hormesis in Aging', *Ageing Res. Rev.*, 7 (2008), 63–78.
24 Ibid., ivi.
25 'The Science of Healthy Aging'.
26 *Mild Stress and Healthy Aging: Applying Hormesis in Aging Research and Interventions*, ed. by Éric Le Bourg and Suresh I. S. Rattan (Dordrecht: Springer, The Netherlands, 2008).

chronic but mild under-nutrition have been reported for human beings. Furthermore, intermittent fasting has been reported to have beneficial effects on glucose metabolism and neuronal resistance to injury. Although at present there are only a few studies performed which utilize mild stress as a modulator of ageing and longevity, hormesis is a useful experimental approach in biogerontology. [27]

Several dietary components, such as vitamins, antioxidants, trace elements, minerals, ethanol, and even herbicides and pesticides have been shown to have typical hormetic dose response. All such compounds (natural or synthetic), which bring about biologically beneficial effects by acting through one or more pathways of maintenance and repair, and stress response, are termed as hormetins. [28] The hormetic effects of various vitamins and macro- and micro-minerals, including iron, iodine, fluorine, selenium and copper have also been reported. [29] Additionally, the effects of zinc also show a typical hormetic dose response, and its beneficial effects are achieved through alterations induced by stress response, in gene expression of various maintenance and repair pathways. [30] Components of various medicinal plants used frequently in the Chinese medicine and in the Indian Ayurvedic system of medicine are claimed to have anti-ageing effects, which appear to be achieved through hormetic pathways. Hormesis may also be an explanation for the health beneficial effects of numerous other foods and food components, such as garlic, Gingko, and other fruits and vegetables. [31] Understanding the hormetic and interactive mode of action of natural and processed foods is a challenging field of research, and has great potential in developing nutritional and other life style modifications for ageing intervention and therapies. For example, it may be possible to develop multi-hormetin formulations as healthy-ageing nutriceuticals whose mode of action is through hormetic pathways by mild stress-induced stimulation of homeodynamic processes.

The last factor for achieving eternal youth is the factor C in the formula $E = GMC^2$. While the G, genes, and M, milieu, components of the formula are being taken care of by various experimental approaches discussed above, the third factor C represents chance, which is the probability of stochastic events leading to a cascade of error-catastrophe in complex interacting systems. Recent de-

27 Ibid., ivi.

28 'The Science of Healthy Aging'; Rehab E. Ali, Suresh I. S. Rattan, 'Curcumin's Biphasic Hormetic Response on Proteasome Activity and Heat Shock Protein Synthesis in Human Keratinocytes', *Ann. NY Acad. Sci*, 1067 (2006), 394–99.

29 Dan P. Hayes, 'Nutritional Hormesis', *Eur J Clin Nutr*, 61 (2007), 147–59.

30 'The Science of Healthy Aging'.

31 'Nutritional Hormesis'; Carlos K. B. Ferrari, 'Functional Foods, Herbs and Neutraceuticals: towards Biochemical Mechanisms of Healthy Aging', *Biogerontology*, 5 (2004), 275–89.

velopments in our understanding of complex networks at all levels of organization, from molecular to societal and global networks, have highlighted the vulnerability of all strong and weak links. Furthermore, this realization of complex networks has reasserted the significance of chance events, which are not amenable to regulation and manipulation.

Thus, understanding the evolutionary basis for the origin of ageing mechanisms as being emergent due to the imperfections of survival mechanisms helps us to reclaim the control of our bodies' homeodynamics and health span. It also helps us to understand that from the nature's side there is neither a mechanism to kill us, nor is there any processes to make us immortal. Therefore, whereas the prospects for achieving healthy ageing and for extending health span are based in solid scientific footing, and will become more and more effective, the dreams of eternal youth are best left to thrive in the domain of the unscientific.

Alessandro Achilli / Ugo A. Perego

Mitochondrial DNA: A Female Perspective in Recent Human Origin and Evolution

Even though modern humans (*Homo sapiens*) have colonized almost every corner of the planet, they are characterized by a very low intraspecies genetic variation, which approximates one mutation per 1,000 base pairs (bps). However, these minimal differences between populations can be utilized to address the challenging questions of how and when the events that lead to their evolution and expansion originated. Over the past 20 years, genetic research has supplemented our knowledge about the origin of modern humans particularly thanks to data generated from mitochondrial DNA (mtDNA) [1] and Y chromosome DNA analysis. [2] The distinctive potential of these two uniparental markers in the study of human past was eventually termed 'archaeogenetics' by Amorim. [3]

Comparing genetic variation between and within populations as it will be explained in great details in this paper is a powerful tool to reconstruct population dynamics that occurred in ancient and modern times. However, DNA alone is just one piece of the puzzle and not the full picture. Geneticists often need to calibrate DNA evidence using data from other scientific fields, such as archaeology, anthropology, linguistics, and even fossil records. It is a comparative process wherein genetic evidence must be studied within a much larger context and not singled out. Scientists working with DNA are faced with the continue challenge of harmonizing their findings with those of other disciplines and when this happens, human history is truly unveiled.

1 Hans-Jürgen Bandelt, Vincent Macaulay, and Martin Richards, *Human Mitochondrial DNA and the Evolution of Homo Sapiens* (Berlin – Heidelberg: Springer Verlag, 2006).
2 Peter A. Underhill and Toomas Kivisild, 'Use of Y Chromosome and Mitochondrial DNA Population Structure in Tracing Human Migrations', *Annual Review of Genetics*, 41 (2007), 539 – 64.
3 Antonio Amorim, 'Archaeogenetics', *Journal of Iberian Archaeology*, 1 (1999), 15 – 25.

I. Why MtDNA in Decoding the Origins?

Two characteristics of the short mitochondrial genome (only 16,569 bps) make it particularly useful for evolutionary and population studies: its mutation rate is much higher than nuclear DNA and, being maternally transmitted, does not undergo recombination. Consequently, while populations dispersed into the different geographical regions of the world, their mtDNAs have been – and still are – accumulating a substantial number of mutations along the radiating maternal lineages. Over the course of time, this process of molecular divergence resulted in monophyletic units (clades) called haplogroups, which are restricted to particular geographic areas or populations. In other words, a haplogroup can be defined as a group of mtDNAs that share a similar combination of mutations (haplotypes) derived from a common female ancestor. The frequency distribution, variation, and coalescence ages of each haplogroup can be estimated by using the 'phylogeographic' approach; then by mapping the mitochondrial lineages in time and space it is possible to make inferences about the demographic history of populations, such as origin, dispersals, range expansion, or migrations. Therefore, by tracking down the origin of mitochondrial DNA, we are in reality reconstructing the history and the migration of the female ancestor of our species.

The complete human mitochondrial genome for a single individual – known as the Cambridge reference sequence (CRS) – was first published in 1981 [4] and later revised in 1999 by Andrews and colleagues. [5] The earliest mitochondrial DNA research to highlight human diversity began by digesting the entire molecule with just a few restriction enzymes. In the early Nineties, the first large-scale population studies were performed on Native Americans, focusing on ancestral migrations from Asia. [6] It was in this specific context that the universally accepted mtDNA nomenclature was initially introduced with the definition of the first four basal (or primary) branches in the phylogenetic tree, named alphabetically as A, B, C, and D. [7] Later these analyses were applied to other continental populations and the cladistic rules for the hierarchical ordering of haplogroups and sub-haplogroups were established. Since then, the naming of new haplogroups has naturally evolved in a self-organizing way, ushered for-

4 Stephen Anderson and others, 'Sequence and Organization of the Human Mitochondrial Genome', *Nature*, 290 (1981), 457–65.

5 Richard M. Andrews and others, 'Reanalysis and Revision of the Cambridge Reference Sequence for Human Mitochondrial DNA', *Nature Genetics*, 23 (1999), 147.

6 Antonio Torroni and others, 'Asian Affinities and Continental Radiation of the Four Founding Native American mtDNAs', *American Journal of Human Genetics*, 53 (1993), 563–90.

7 These four pan-American mtDNA lineages are today called A2, B2, C1, D1 to distinguish them from their Asian counterparts (see Figure 1).

ward by those who produced novel data, following the rules of the nomenclature system and respecting the published record. Today, an ever-increasing number of new complete sequences are becoming available, resulting in a more refined picture of the mtDNA world phylogeny [Figure 1] and the achievement of greater precision in terms of timing and identification of ancient and recent human dispersals [Figure 2].

III. The Origin of Modern Humans

By the 20th century, the debate about the origin of modern *H. sapiens* was still unsolved and two opposite scenarios were proposed. According to the first, the distribution of anatomical traits in modern human populations in different regions was inherited from local populations of *Homo erectus* and intermediate 'archaic' forms. This 'Multiregional Hypothesis' states that all modern humans evolved in parallel from earlier populations in Africa, Europe, and Asia, with some genetic intermixing among these regions. [8] Support for this thesis came from the similarity of certain minor anatomical structures in modern human populations and preceding populations of *H. erectus* in the same regions. A different model proposed that a small, relatively isolated population of early humans evolved into modern *H. sapiens*, and that this population succeeded in spreading out of Africa, towards Europe and Asia, replacing all of the Neanderthals and other late archaic *H. sapiens*. [9] Support for this theory was provided by archaeological data, since the earliest modern *H. sapiens* skeletons came from Africa. If this interpretation of the fossil record is correct, all people today share a relatively recent African ancestry and the variation among modern populations is a recent phenomenon. This hypothesis is also referred to as the 'out of Africa' model.

One of the best instruments to resolve the debate about human origin is mtDNA. In 1987, a pioneering study performed by Wilson's research group launched their support to the out of Africa theory. [10] They refocused the debate around the opposition between a recent African origin *versus* multiregional evolution.

8 Alan G. Thorne and Milford H. Wolpoff, 'Regional Continuity in Australasian Pleistocene Hominid Evolution', *American Journal of Physical Anthropology*, 55 (1981), 337–49.

9 Chris B. Stringer and Peter Andrews, 'Genetic and Fossil Evidence for the Origin of Modern Humans', *Science*, 295 (1988), 1263–68.

10 Rebecca L. Cann, Mark Stoneking, and Allan C. Wilson, 'Mitochondrial DNA and Human Evolution', *Nature*, 325 (1987), 31–36.

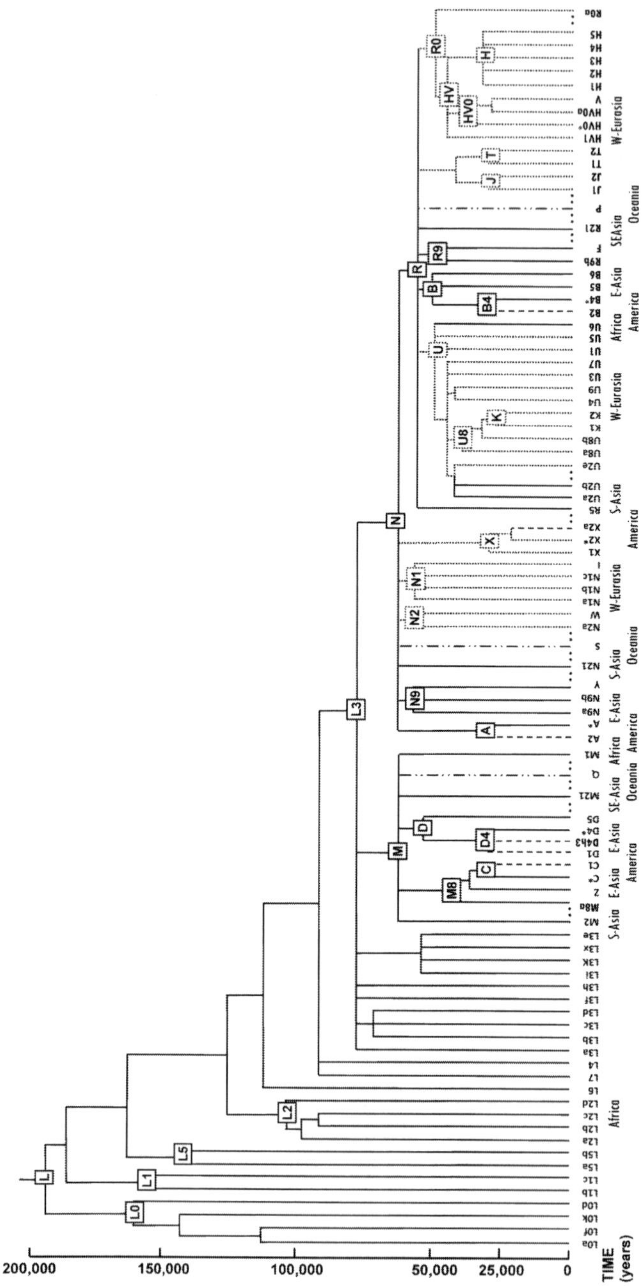

Figure 1: Schematic phylogeny of human mtDNA

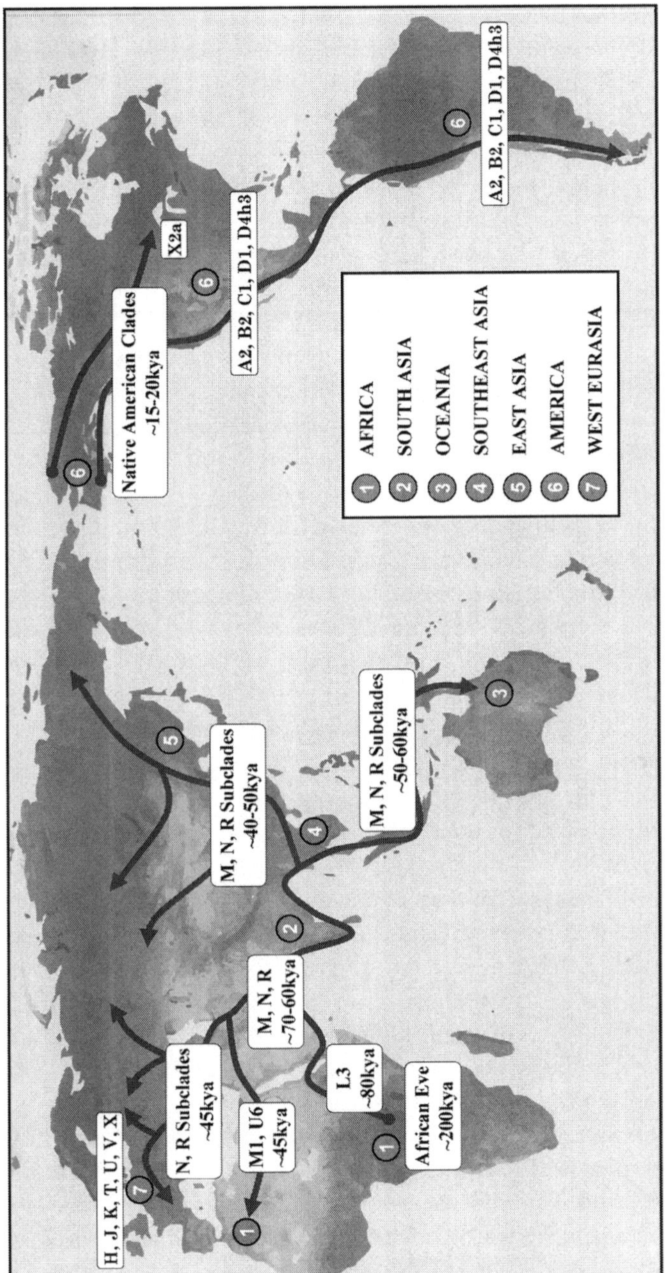

Figure 2: Human migration routes based on mtDNA data

In such work, the researchers extracted mtDNA from 147 individuals, drawn from five geographic populations (Africans, Asians, Caucasians, aboriginal Australians, and aboriginal New Guineans) and built a phylogenetic tree by connecting all the lineages and dating them. The maternal genealogy was divided into two basal branches: one of these clades was found only in African-Americans, whereas the other was found in African-Americans and throughout the other populations sampled. This pattern implied that all mitochondrial DNAs stem from one woman who lived about 200,000 years ago somewhere in Africa, which was later dubbed as the 'African Eve'. The 'Eve' in question was actually the most recent common ancestor through the matrilineal descent of all humans living today. Therefore mtDNA phylogeography played a pivotal role in supporting a single origin for all modern humans, by anchoring our most recent (maternal) common ancestor (MRCA) to sub-Saharan Africa and suggesting a dispersal wave which populated the rest of the world from this continent. [11]

Knowing that we are all descendants from ancient African progenitors might still be a strange thing to accept by some. Obviously, the first thing coming to the mind is the large variation in phenotypes observed between modern-day Africans and other populations. As explained in the introduction of this paper, we already emphasized the close genetic resemblance found in every living person (one difference every 1,000 bases), meaning that from a DNA point of view, we are all more similar than we are different. Of course, genetic mutations combined with environmental adaption in the millennia that characterized the human colonization of the globe are at the source of the outward distinctions detected in people from different ethnic groups today. With regards to the present review, the African origin of all mitochondrial lineages has to be considered in the context of the large number of ancestors each person had. In other words, although we are all descendants of 'African Eve,' her matriarchal legacy survived almost exclusively in the mitochondrial genome. Considering that the number of potential ancestors double at each generation, every person would have had over one billion progenitors just thirty generations ago (~750/1,000 years ago). Of all these ancestors, only the unbroken umbilical lineage is what ties each one of us back to the 'African Eve', while the record written in the nuclear DNA is the result of the recombination from the million of ancestors we all share.

The debate about how many distinct dispersals of modern humans contributed to the modern pool of *H. sapiens* and which routes were taken across Eurasia is still under investigation. This debate was fuelled by the finding of some modern human remains in the Near East (Skhul & Qafzeh) that were dated

11 Antonio Torroni and others, 'Harvesting the Fruit of the mtDNA Tree', *Trends in Genetics*, 22 (2006), 339–45.

around 90 – 120,000 years ago. [12] These ages preceded the widespread appearance of the Upper Palaeolithic in Eurasia about 50,000 years ago and the appearance of modern humans in Australia, previously estimated around 70,000 years ago. [13] One possibility was that the Skhul and Qafzeh remains were the traces of an earlier dispersal than the one which gave rise to the Upper Palaeolithic of Eurasia. If this hypothesis is true, this early diffusion event was very localized and short-lived, leaving no descendants in modern human populations. [14]

As for the contribution to the present-day gene pool, two main expansion hypotheses of modern humans out of Africa have been proposed. According to the 'Levantine corridor' hypothesis, the initial spread occurred via North Africa and the Nile valley, with subsequent dispersals to both the West into Europe and to the East into Asia. According to the 'southern' or 'coastal' route hypothesis, the initial dispersal was from Ethiopia, across the mouth of the Red Sea, and then either northward through Arabia or Eastward along the south Asian coastline to Australasia.

Genetic evidences, together with archaeological and palaeo-anthropological records, were in favour of the Levantine corridor route. [15] However, it has been suggested that an extant tropical belt of human populations anatomically resembling sub-Saharan Africans (with 'racial' features such as very dark skin and curly hair) might represent the survivors of the southern coastal route. [16] They include some Southern Indians, the Andamans, the so-called Negritos of the Philippines and the Malay Peninsula, Papuans and Aboriginal Australians. As for mtDNA data, in 1997 Watson and colleagues provided evidence for a single dispersal from Africa (although not distinguishing between a northern or southern route). [17] Further mtDNA analyses (on haplogroup M) supported an early southern migration from Africa to India and beyond ~60,000 years ago, leaving open the possibility for a later northern course into the rest of Eurasia. [18]

---

12 Chris B. Stringer and Rainer Grün, 'Palaeoanthropology. Time for the Last Neanderthals', *Nature*, 351 (1991), 701 – 02.

13 Alan G. Thorne and others, 'Australia's Oldest Human Remains: Age of the Lake Mungo 3 Skeleton', *Journal of Human Evolution*, 36 (1999), 591 – 612.

14 Peter Mellars, 'Why Did Modern Human Populations Disperse from Africa ca. 60,000 years ago? A New Model', *Proceedings of the National Academy of Sciences of the United States of America*, 103 (2006), 9381 – 86.

15 Luigi L. Cavalli-Sforza, Paolo Menozzi, and Alberto Piazza, *The History and Geography of Human Genes* (Princeton: Princeton University Press, 1994).

16 Masatoshi Nei and Arun K. Roychoudhury, 'Evolutionary Relationships of Human Populations on a Global Scale', *Molecular Biology and Evolution*, 10 (1993), 927 – 43.

17 Elizabeth Watson and others, 'Mitochondrial Footprints of Human Expansions in Africa', *American Journal of Human Genetics*, 61 (1997), 691 – 704.

18 Lluís Quintana-Murci and others, 'Genetic Evidence of an Early Exit of *Homo Sapiens Sapiens* from Africa through Eastern Africa', *Nature Genetics*, 23 (1999), 437 – 41.

Yet, some geneticists rejected the existence of a southern migration and the question of a single versus multiple dispersals is still disputed. [19]

The current mtDNA phylogeny, based now on a large number of complete sequences (Figure 1) has been critical to resolving the debate. The deepest and most ancient branches in the phylogeny are all localized in Africa, while all non-African haplogroups encompass variants of a single mtDNA clade, haplogroup L3, whose derivatives are found both in Africa and out of Africa. L3 traces back to sub-Saharan Africa, most probably East Africa arising about 83,000 years ago. This situation is well aligned with a single exodus from Africa that occurred at a time bracketed between the occurrence of haplogroup L3 and the appearance of the two basal non-African clades, N and M (dated ~63,000 years old), followed closely by the appearance of lineage R (diverging rapidly within haplogroup N ~60,000 years ago). These data support a single out of Africa exit event between 80 – 85,000 and 60 – 65,000 years ago. Actually, the northern route along the Levantine corridor, even if apparently easily accessible from East Africa, did not explain why adjacent Europe was settled thousands of years later than distant Australia. [20] Two papers published in *Science* in 2005 addressed this matter. Macaulay's research team analysed the Orang Asli, the aboriginal inhabitants of the Malay Peninsula, [21] while Thangaraj and colleagues focused on the Andaman islanders, called 'Negritos' (for the characteristic phenotype of dark skin), [22] both groups performing a large number of complete mitochondrial sequences in order to clarify the origin of these populations. They discovered that both Orang Asli and Andaman islanders harboured ancient mtDNA lineages, belonging to the founder haplogroups M, N, and R, with coalescence ages of ~44,000 to ~63,000 years, which were considered the legacy of an early diffusion of modern humans out of Africa. Thus, there was a single rapid out of Africa dispersal (~70,000 years ago) involving a founding group of individuals harbouring the L3 mtDNA haplogroup and starting from the Horn of Africa towards the Persian Gulf and further along the tropical coast of the Indian Ocean to Southeast Asia and Australasia. During this coastal migration, haplogroups M, N, and R evolved and the ancestral L3 was lost. Moreover, this scenario is strongly supported by palaeoenvironmental evidence, confirming that a northern migration would

19 Richard Cordaux and Mark Stoneking, 'South Asia, the Andamanese, and the Genetic Evidence for an 'Early' Human Dispersal Out of Africa', *American Journal of Human Genetics*, 72 (2003), 1586 – 90.

20 Peter Forster and Shuichi Matsumura, 'Evolution. Did Early Humans Go North or South?', *Science*, 308 (2005), 365 – 66.

21 Vincent Macaulay and others, 'Single, Rapid Coastal Settlement of Asia Revealed by Analysis of Complete Mitochondrial Genomes', *Science*, 308 (2005), 1034 – 36.

22 Kumarasamy Thangaraj and others, 'Reconstructing the Origin of Andaman Islanders', *Science*, 308 (2005), 996.

have been impossible during the glacial period extending from ~70,000 to 50,000 years ago. Descendants of all three macrohaplogroups M, N, and R can be found today all over Eurasia, America, Australasia, and Oceania, with only few uncommon regional exceptions.

Soon after the rapid expansion along the coastlines of Southern Asia, South-Eastern Asia, and Indonesia, the wave of migration reached New Guinea and Australia, at a time when the lower sea levels joined the two islands into one land mass, with sea travel limited to narrow straits. [23] The deep and specific phylogenetic lineages indicate a small founding population size followed by isolation of Australian and, to a lesser extent, New Guinea aboriginal populations. Additionally, this separation could explain the morphological changes observed among Australian fossil records and could also help in understanding the remarkably restricted range of Pleistocene Australian lithic industries and bone artefacts as compared to contemporary cultures elsewhere in the world. [24]

The richest basal variation in the founder haplogroups M, N, and R is found along the southern stretch of Eurasia, particularly in the Indian subcontinent (Figure 1), suggesting a rapid colonization along the southern coast of Asia, reaching Sahul ~60,000 years ago. The expansion northwards to fill the heartland of the continent took place at a later time (~45,000 years ago) when a combination of technology and climatic conditions enabled the exploration of Eurasia inland.

Western Eurasians, in contrast with Southern Asians, Eastern Eurasians, and Australasians, have a high level of haplogroup diversity within haplogroup N and R, but lack haplogroup M almost entirely (Figure 1). Thus, the colonization of West Eurasia is thought to have been the result of an offshoot of the colonization along the southern route, followed by a lengthy pause, perhaps around the Persian Gulf, until the climate improved and humans were able to enter Europe from the Levant. This conclusion is further supported by the relatively late arrival of modern humans into Europe (~46 – 41,000 years ago), which is also confirmed by archaeological evidence. [25]

Although haplogroup M differentiated soon after the out of Africa exit and it is widely distributed in Asia (East Asia and India) and Oceania, there is an interesting exception for one of its more than 40 sub-clades: M1 (black in Figure

23 Georgi Hudjashov and others, 'Revealing the Prehistoric Settlement of Australia by Y chromosome and mtDNA Analysis', *Proceedings of the National Academy of Sciences of the United States of America*, 104 (2007), 8726 – 30.

24 Peter Mellars, 'Going East: New Genetic and Archaeological Perspectives on the Modern Human Colonization of Eurasia', *Science*, 313 (2006), 796 – 800.

25 Peter Mellars, 'A New Radiocarbon Revolution and the Dispersal of Modern Humans in Eurasia', *Nature*, 439 (2006), 931 – 35.

1). Indeed, this lineage is mainly limited to the African continent with peaks in the Horn of Africa.

There are two potential scenarios that could explain the peculiar geographical distribution of M1. First, it is possible that the entire M super-haplogroup could have differentiated in East Africa before the out of Africa exit and M1 could be a remnant of the original African M haplogroup. Alternatively, M1 could be the only African representative of M, in contrast to the numerous sister-clades found in Asia and Oceania, indicating a likely back-migration into East Africa.

The scenario of a back-migration into Africa seems to be also supported by haplogroup U6 (black in Figure 1). Haplogroup M's Eurasian sister clade, haplogroup N, has a very similar age to M and no indication of an African origin. Haplogroup R diverged rapidly within haplogroup N ~60,000 years ago and in turn embraces U branch (Figure 1). Haplogroup U is characterized by an extremely broad geographical distribution ranging from Europe to India and Central Asia, differentiating in Western Eurasia around 59,000 years ago, and is subdivided into numerous clades (currently U1-U9). However, the clade U6 is mainly found in Northern Africans with the highest frequency in Eastern Africa. Based on the recent work of Olivieri and colleagues, it is plausible to conclude that there was a gene relocation event about 40 – 45,000 years ago that resulted in a return to Africa of modern humans through the Levantine and it is witnessed by today's presence of haplogroups M1 and U6 in Northern Africa. [26]

III. The Origin of the Europeans

The first modern human populations diffused across Europe are known as Cro-Magnons (named after the cave in Southwest France where the first remains were found in 1868). They were hunter-gatherers and probably originated in the Middle East where similar specimens came to light.

At the time the Cro-Magnons populated Europe, the continent was occupied by a different species of the *Homo* genus, named *Homo neanderthalensis*. Neanderthals were one group of archaic hominids that dominated the European scene for about 200,000 years, until their abrupt extinction between 30 – 40,000 years ago, being replaced by anatomically modern humans. Although the causes of their extinction have not yet been fully explained, there is plenty of evidence to support that the two species coexisted in Europe for ~6,000 years, even to the extent of sharing common areas (e.g. Châtelperron, central France) for perhaps

26 Anna Olivieri and others, 'The mtDNA Legacy of the Levantine Early Upper Palaeolithic in Africa', *Science*, 314 (2006), 1767 – 70.

1–2,000 years.[27] One question that was answered only recently through DNA was whether or not Neanderthals contributed to the genetic pool of modern humans. Given their smaller size relative to the nuclear genome and the typical presence of several hundred molecules in each cell, mtDNA from ancient remains provided important clues into this matter. In 1999, the research group directed by Svante Pääbo published a 360 bp long mtDNA sequence of the HVS-II region of a Neanderthal specimen, found in 1856 in central Europe (Feldhofer).[28] They discovered that the divergence of Neanderthal mtDNA from the line leading to the contemporary human mtDNA gene pool is almost three times higher ('older' in terms of time) than the deepest divergence among contemporary human mtDNAs. These data showed that Neanderthal and Sapiens ancestral mtDNA gene pool evolved independently for a substantial period of time and that there was no apparent interbreeding between the two species.

More recently, thanks to the improvement in high-throughput sequencing techniques, a 65,250 bp long sequence of Neanderthal nuclear genome was also published,[29] followed by over one million nuclear base pairs recovered from the same ancient sample[30] and the first complete mitochondrial genome sequence.[31] The conclusions reached by the different research teams working on Neanderthal DNA were basically the same: Neanderthal DNA (both mitochondrial and nuclear) fall outside the variation of extant human DNA, thus implying that Neanderthals did not contribute to the origin of modern humans.

As previously reported, the rapid spread of the Cro-Magnons (characterized by the appearance of Aurignacian industry) is thought to have been the result of an offshoot of the main dispersal along the southern route, probably facilitated by the climate improvement, which led the ancestors of West Eurasians to enter first the Levant and then Europe. On the basis of discoveries of fossilized human skeletal remains and a number of long stratified archaeological sequences in Israel and Lebanon, it has been recognized that populations of anatomically modern humans, equipped with typically Upper Palaeolithic technology (proto-

27 Brad Gravina, Peter Mellars, and Christopher Bronk Ramsey, 'Radiocarbon Dating of Interstratified Neanderthal and Early Modern Human Occupations at the Chatelperronian Type-site', *Nature*, 438 (2005), 51–56; Id., 'Confirmation of Neanderthal/Modern Human Interstratification at the Chatelperronian Type-site', *Proceedings of the National Academy of Sciences of the United States of America*, 104 (2007), 3657–62.

28 Matthias Krings and others, 'DNA Sequence of the Mitochondrial Hypervariable Region II from the Neanderthal Type Specimen', *Proceedings of the National Academy of Sciences of the United States of America*, 96 (1999), 5581–85.

29 James P. Noonan and others, 'Sequencing and Analysis of Neanderthal Genomic DNA', *Science*, 314 (2006), 1113–18.

30 Richard E. Green and others, 'Analysis of One Million Base Pairs of Neanderthal DNA', *Nature*, 444 (2006), 330–36.

31 Richard E. Green and others, 'A Complete Neanderthal Mitochondrial Genome Sequence Determined by High-Throughput Sequencing', *Cell*, 134 (2008), 416–26.

Aurignacian) appeared in the ancient Near East by ~47 – 49,000 years ago, and had apparently spread to parts of South-Eastern Europe shortly after this time. [32]

The term Palaeolithic was coined by archaeologist John Lubbock in 1865, and refers to a Prehistoric era distinguished by the development of the first stone tools. Palaeolithic is traditionally divided into three periods: the Lower Palaeolithic (from ~2.6 million to 300,000 years ago), the Middle Palaeolithic (~300 to 30,000 years ago) and the Upper Palaeolithic (~50 to 11,500 years ago). Thus, the Upper Palaeolithic encompasses most of the events that led to the occupation of the Old World by modern humans. However, the archaeological term 'Upper Palaeolithic' is often used when referring to remains such as stone or bone tools that represent cultures, not time periods. During the Upper Palaeolithic, the observed cultural and technological characteristics, as well as the population increase, were more rapid and had distinct global effects across Eurasia and Africa, when compared with the slow pace of cultural changes during the Middle Palaeolithic. Indeed, after the appearance of anatomically modern humans in Africa, around 200,000 years ago, the following period of about 60,000 years saw a change in the lifestyle and cultural features of humans from their predecessors. The Middle to Upper Palaeolithic transition outside of Africa is sometimes referred to as the 'Upper Palaeolithic revolution' because of the rapid and extensive nature of changes that took place in some regions. [33] Flake-based industries were replaced by blades, bone tools, and perforated ornaments became common. Long-distance trading was also established during that era. The Upper Palaeolithic revolution process that most likely began in a core area and expanded by demic-diffusion migration over long distances across Africa and Eurasia (except for Southeast Asia) is still a matter of intense study. [34]

In the late Upper Palaeolithic, between ~25 and ~19.500 years ago, Europe witnessed the Last Glacial Maximum (LGM). The sea fell to as much as 120 meters below the present level. Northern and Central Europe was largely covered with glaciers or polar desert, while the South was characterized by semi-desert or grassland. These hostile conditions modified the distribution of plants and animals, and therefore humans. Humans were forced to retreat southwards to the warmest regions, or glacial refugia. The major refuge areas were localized respectively in south-western France/Cantabria (Atlantic and Western Mediterra-

32 Janusz Kozlowski and Marcel Otte, 'The Formation of the Aurignacian', *Journal of Anthropology Research*, 56 (2000), 513 – 24.

33 Ofer Bar-Yosef, 'The Upper Paleolithic Revolution', *Annual Review of Anthropology*, 31 (2002), 363 – 93.

34 Mark A. Jobling, Matthew E. Hurles and Chris Tyler-Smith, *Human Evolutionary Genetics: Origins, Peoples and Disease* (New York: Garland Science, 2004), pp. 285 – 87.

nean zone), in Italy near the Apennines, in the Balkans, and in Ukraine/Central Russian Plain (Periglacial zone). [35]

Humans remained secluded in these areas until climatic conditions improved starting about 13,000 years ago. This event was probably followed by local re-expansions from about 11,500 years ago after the relapse to glacial conditions in the last phase of the glaciation, called Younger Dryas. The end of the LGM was followed by relative climatic stability during which Mesolithic human populations prospered and grew in size, and where agriculture was introduced in Europe by farmers from the Middle East, marking the beginning of the Neolithic age.

The term Neolithic (10 – 4,000 years ago) is used, especially in archaeology and anthropology, to designate a stage of cultural evolution or technological development characterized by the use of stone tools, the existence of settled villages largely dependent on domesticated plants and animals, and the presence of crafts such as pottery and weaving. The Neolithic appeared at different times in different regions, corresponding with the first appearance of agriculture. The advent of agriculture usually distinguishes the Neolithic from earlier Palaeolithic or Mesolithic hunting, fishing, and food-gathering cultures. The hunter-gatherer way of life was replaced by domestication of crops and animals, enabling people to live more sedentary lives. Permanent settlements arose, creating new social, cultural, economic and political institutions.

The Neolithic period of interest for Europe began around 10,000 years ago in the Near East, [36] more specifically in the most western area of the Fertile Crescent, in the south-eastern and south-central part of Asia Minor. Its rate of expansion was uniform and fast, covering about one kilometre per year; indeed the entire process, from Greece to British Isles, took place in about 2,500 years and was characterised by bronze utensils and weaponry, though stone technology was still used.

There are two models to explain the spread of agriculture. In the 'cultural diffusion' model, the farmers did not move, but the technology and the ideas did. The second model involves gene flow, in which the farmers themselves moved, taking agricultural practice with them. While there are many different gene-flow models, the most discussed is 'demic diffusion', also known as 'wave of advance', proposed by Ammerman and Cavalli-Sforza in 1984, also known as 'colonization without colonists'. [37] In this model, spread is stimulated by population growth (itself a result of increasing availability of food) and local migratory activity. In a

35 Marcel Otte, 'The Northwestern European Plain around 18.000 BP', in *The World at 18.000 BP*, ed. by Olga Soffer and Clive Gamble (London: Unwin, 1990), pp. 54 – 68.

36 Followed by the Bronze Age which (~4,000 years ago) also spread into Europe from the East.

37 Albert J. Ammerman and Luigi L. Cavalli-Sforza, *The Neolithic Transition and the Genetics of Populations in Europe* (Princeton: Princeton University Press, 1984).

pure acculturation model the spread of farmers should not be associated with major genetic changes. In a pure demic diffusion model without interbreeding the migration of farmers would have the effect of replacing the gene pool of indigenous Europeans with that of the Near Eastern immigrants. The debate that has arisen around the genetic evidence is based not upon either extreme position, which nobody accepts as reasonable, but on mixed intermediate models and on their consequences for modern European genetic diversity.

In summary, the genetic composition of modern European populations has been influenced by three major demographic events: a) the settlement of hunter-gatherers during the Upper Palaeolithic (~40 – 45,000 years ago); b) the later re-expansion (~11 – 13,000 years ago) after the Last Glacial Maximum from glacial refugia; c) the arrival of the Neolithic farmers who spread from the Near East (~10,000 years ago). Archaeologists, anthropologists, linguists, and more recently geneticists have attempted to evaluate the contribution of these three events, in a still open debate on what can be defined as the 'Neolithic *versus* Palaeolithic legacy'.

Through a pioneering work, Cavalli-Sforza and colleagues analysed classic genetic markers (blood groups, tissue antigens, various enzymes) in 26 European and Near Eastern populations and concluded that the European genetic landscape bore the imprint of a succession of expansion waves, with a predominant role of the Neolithic colonization (the already mentioned 'demic diffusion' hypothesis).[38] But this was not the end of the story. Depending on the markers and the approaches used, the Neolithic contribution has been estimated, in a range from 100 %,[39] less than 25 % or less than 1 % (on the basis of ancient mtDNA data).[40]

Early mtDNA studies of European populations revealed that all Europeans essentially share the same set of haplogroups as Near Easterners and that those haplogroups were absent from Sub-Saharan Africans and East Asians.[41] These observations confirmed that Europeans and Near Easterners share a rather recent common ancestry. From the complete mtDNA phylogeny that is now available (Figure 1), we can see that European mtDNA variation – embedded as it is in the West Eurasian mtDNA variation – is surprisingly impoverished in the number of independent basal lineages in comparison with South Asian mtDNA

38 Paolo Menozzi, Alberto Piazza, and Luigi L. Cavalli-Sforza, 'Synthetic Maps of Human Gene Frequencies in Europeans', *Science*, 201 (1978), 786 – 92.

39 Isabelle Dupanloup and others, 'Estimating the Impact of Prehistoric Admixture on the Genome of Europeans', *Molecular Biology and Evolution*, 21 (2004), 1361 – 72.

40 Wolfgang Haak and others, 'Ancient DNA from the First European Farmers in 7500-year-old Neolithic Sites', *Science*, 310 (2005), 1016 – 18.

41 Antonio Torroni and others, 'Classification of European mtDNAs from an Analysis of Three European Populations', *Genetics*, 144 (1996), 1835 – 50.

variation. The bulk of European mtDNA lineages stem from just three R branches, haplogroups R0, JT, and U, as well as three minor N branches, N1, N2, and X. This reduction in basal variation may be best explained by the peripheral role that the pioneer migration into the Near East and subsequently Europe must have played in the broader out of Africa scenario (in contrast, by the way, to traditional accounts which gave the Near East a pivotal role). The variation in the basal European mtDNA haplogroups can be dated to about 25,000 – 50,000 years (Figure 1). This, however, does not yet clarify to what extent the first migration of modern humans into Europe shaped the mtDNA pool of modern Europeans, since later immigration from the Near East could have replaced the descendants of the first settlers. The wisdom received in the Nineties poses that the European gene pool predominantly resulted from Near Eastern newcomers bringing the Neolithic culture with them. The new suggestion was that only a small minority of lineages dated to Neolithic immigration, while the remaining lineages – dating back to between 15,000 and 50,000 years ago – seemed to have autochthonous roots of Early, Middle or Late Upper Palaeolithic origin within Europe. The results were, however, rather tentative because they relied on comparisons with a very small and inadequate sample from the Near East. Later studies provided much stronger evidence showing that more than three-quarters of present-day European mtDNAs most likely stemmed directly from indigenous Palaeolithic ancestors. Additionally, with a sufficiently large sample size and a better resolved phylogeny, these studies showed putative clades of mtDNA exhibiting gradients similar to those of other marker systems. [42]

The dissection of one minor autochthonous European mtDNA clade (haplogroup V) and the subsequent analysis of its geographic distribution constituted the first clear evidence that late-glacial expansions of Palaeolithic populations from refuge areas, in particular localized in South-Western Europe (Franco-Cantabrian refuge area) could have had a major impact in repopulating the continent. [43] Further evidence of the overwhelming importance of the Franco-Cantabrian glacial refuge has been provided afterwards by dissecting haplogroup H, which is the most common haplogroup in Europe (40 %-50 % of mtDNAs), and haplogroup U into specific sub-lineages. [44] The age estimates and geographic distributions of V, as well as that of sub-haplogroups H1, H3, and U5b (the latter explained in more details below) strongly support a repopulation

42 Martin Richards and others, 'In Search of Geographical Patterns in European Mitochondrial DNA', *American Journal of Human Genetics*, 71 (2002), 1168 – 74.

43 Antonio Torroni and others, 'A Signal, from Human mtDNA, of Postglacial Recolonization in Europe', *American Journal of Human Genetics*, 69 (2001), 844 – 52.

44 Alessandro Achilli and others, 'The Molecular Dissection of mtDNA Haplogroup H Confirms that the Franco-Cantabrian Glacial Refuge was a Major Source for the European Gene Pool', *American Journal of Human Genetics*, 75 (2004), 910 – 18.

of much of Western and Northern Europe from the south-western refuge in the Late Upper Palaeolithic. Molecular evidence using a mitochondrial DNA clock indicates that the majority of extant European lineages (analysed so far) coalesce to a small number of founders approximately 13 – 16,000 years ago. This is most likely the effect of a population bottleneck followed by re-expansion, which the distribution of haplogroups H1, H3, V and U5b suggests to have occurred in Southwest Europe.

Regarding haplogroup U5b, an unexpected genetic link was observed in a study headed by Achilli and his collaborators in 2005,[45] where seven new complete mtDNA sequences (one Berber from Algeria, two from Italy, one Spanish, and three Saami) clustered into U5b1b, a sub-clade encompassing also the Yakut and Fulbe mtDNAs, and having a coalescent age of ~8,600 years ago. The recent common female ancestor found in populations living as far as 9,000 miles apart and with distinctive anthropological affinities was considered an unequivocal link between Berbers and Saami. The plausible hypothesis from these data is that their common ancestors lived also in the Franco-Cantabrian refuge area during the Last Glacial Maximum and indicates a double migration route that the European hunter-gatherers followed: northward toward the Scandinavian Peninsula and southward by crossing the Strait of Gibraltar, bringing the U5b1b lineage to modern North Africans.

IV. The Origin of the Etruscans

From Africa to Eurasia going through Palaeolithic, LGM and Neolithic events, we can conclude the journey of man from a mtDNA perspective by providing a classic example of relatively recent history involving one of the most mysterious European civilizations: the Etruscans.

The origin of the Etruscan people has been a source of controversy for the past 2,500 years. Their sophisticated culture developed and prospered in the 9th century BCE over a region of central Italy named Etruria that became dominant in the Central Mediterranean area. However, in the year 509 BCE, the last Etruscan king of Rome was removed from power and the Roman Republic was established. From that moment the expansion period of the Etruscans ended, followed by a slow process of conquest and assimilation by the Romans. All of this culminated with the attribution of Roman citizenship to all Etruscans about 90 – 88 BCE. Rapidly the Etruscan culture and their peculiar language disappeared, possibly because Etruscan lands were often distributed to Roman ve-

45 Alessandro Achilli and others, 'Saami and Berbers – An Unexpected Mitochondrial DNA Link', *American Journal of Human Genetics*, 76 (2005), 883 – 86.

terans and partially repopulated by poor Roman citizens. However, despite a potentially extensive dilution, there is no evidence that the Etruscan gene pool completely disappeared. In addition, the initial assimilation was probably mainly male-mediated, with Roman veterans marrying women from local communities. Thus, it could be expected that some populations of modern Tuscany would have retained at least a fraction of the ancestral gene pool, in particular the maternally-transmitted mtDNA.

In 'Mitochondrial DNA Variation of Modern Tuscans Supports the Near Eastern Origin of Etruscans', published on the April 2007 issue of the *American Journal of Human Genetics*, it is reported how researchers performed a phylogenetic analysis of mtDNAs from modern Tuscan populations in order to provide new clues about the origins of the Etruscans and assess whether or not the extant population still carries that mtDNA signal. [46] The sample collection was represented by 322 subjects from three areas of Tuscany: 86 from Murlo, a rather isolated town of Etruscan origin in the Siena province; 114 from Volterra, a former major Etruscan city in the province of Pisa; and 122 from the Casentino Valley in the province of Arezzo, which was part of historical Etruria. The analysis of the control-region variation in the three populations, followed by a hierarchical survey of haplogroup and sub-haplogroup diagnostic markers in the coding region, revealed that Tuscan mtDNAs clustered into 39 haplogroups (209 haplotypes) and, intriguingly, more than 17% of mtDNAs from Murlo belonged to haplogroups U3, U7, HV and R0a, which are typically found among Near Eastern populations. A comparison with mtDNAs from 55 Western Eurasian populations (including seven from other Italian regions) was carried out to determine whether the overall haplogroup frequencies in the three Tuscan samples were indeed different. This comparison was carried out with a principal-component analysis (PCA), which is a statistical method used to reduce multidimensional data sets (the many haplogroups) to lower dimensions (axes of a chart) and determine the coordinates for the different populations on a map. On one such 'genetic map' Murlo was placed relatively close to Near Eastern populations due to the already-mentioned high frequencies for haplogroups U3, U7, HV and R0a.

In addition, the same comparison revealed that more than 5% of the haplotypes from the Tuscan populations were found in Near Eastern populations, but absent in a total of more than 10,000 mtDNAs from Europe, including more than 2,300 mtDNAs from other Italian regions. In conclusion, the mtDNA data from this study support the scenario of a post-Neolithic direct genetic input

46 Alessandro Achilli and others, 'Mitochondrial DNA Variation of Modern Tuscans Supports the Near Eastern Origin of Etruscans', *American Journal of Human Genetics*, 80 (2007), 759–68.

from the Near East to the present-day population of Tuscany. A scenario that is in agreement with the Anatolian origin of Etruscans, as proposed by several classical historical writers, included Herodotus:

> In the days of Atys, the son of Manes, there was great scarcity through the whole land of Lydia [...]. So the king determined to divide the nation in half, and to make the two portions draw lots, the one to stay, the other to leave the land. He would continue to reign over those whose lot it should be to remain behind; the emigrants should have his son Tyrrhênus for their leader. The lot was cast, and they who had to emigrate went down to Smyrna, and built themselves ships, in which, after they had put on board all needful stores, they sailed away in search of new homes and better sustenance. After sailing past many countries they came to Umbria, where they built cities for themselves, and fixed their residence. Their former name of Lydians they laid aside, and called themselves after the name of the king's son, who led the colony, Tyrrhenians. [47]

Later the Romans called them Etrusci or Tusci.

In conclusion, the study of mtDNA variation in ancient and modern day samples is a paradigmatic example of the power of genetic inference in human origin and evolution. It is now directly fostering the development of new research avenues and providing valuable insights in other fields such as palaeontology, archaeology, and linguistics. The overall objective remains the same: reconstructing the events that characterized human history by adding the previously unavailable genetic dimension.

Acknowledgments

We are grateful to Dr. Anna Olivieri and Dr. Maria Pala for commenting on the manuscript, Ann Turner and Manndi Loertscher for reading a late version of the text and providing valuable feedbacks.

47 Herodotus, *The Histories*, c. 430 BCE, *Book One*, 94, trans. with Notes by George Rawlinson, With an Introduction by Tom Griffith (London: Wordsworth, 1996), pp. 45–46.

Paola Spinozzi

The Origin of Art: Towards Humanistic-Scientific Theories and Methodologies

> Much of what we call achievement in art is implicit in our idea of the origins of a work of art.
> Denis Dutton, 'Authenticity in Art', 2003

I. Beyond Humanistic Approaches to Art and Its Origin

The claim that enquiries into the origin of art demand theories and methodologies at the interface of the humanities and the sciences calls for a thorough reassessment of long-acknowledged, canonical critical orientations. In order to show that osmosis between humanistic and scientific interpretative frameworks can provide novel insights into the concept of origin in art, it is necessary to explore why art has always been regarded as a sphere of investigation reserved for humanists, and why artistic creativity, aesthetic quality, and aesthetic experience have been explored almost exclusively with the aid of categories derived from philosophy and the history of art. My argument is that if humanistic and scientific disciplines agree to explore trans-disciplinary routes, long-established definitions of art can be re-framed and, more specifically, innovative hypotheses about the origin of artistic expression can be developed. Furthermore, becoming acquainted with scientific approaches to the origins of art engenders a meta-critical process, which prompts humanities scholars to ask themselves how their specific research methods can gain greater efficacy.

Trespassing on the carefully guarded disciplinary threshold of art history and aesthetics is an act which exposes wide-ranging methodological questions: what is the impact of scientific frameworks on our understanding of the origin of art? How do scientific studies of the arts affect the theories and critical tools applied by humanists?

The relevance of scientific approaches to art will be explored along three lines: firstly, hermeneutic paradigms long established in aesthetics will be reconsidered; secondly, research methods from the fields of archaeology, anthropology, cognitive sciences and biology will be examined; finally, acquisitions arising from the adoption of critical orientations at the interface of the humanities and the sciences will be discussed.

II. Humanistic and Scientific Paradigms on Art

Humanistic orientations in the study of art entail assumptions tailored to validate and empower the foundations of Western culture. To questions such as 'What purpose does art pursue?' and 'What is the function of art?' certain answers, like 'Beauty is the major purpose of art', or 'The function of art is to represent Truth', have traditionally been offered for the fine arts, and other answers, such as 'Art achieves instrumental ends', have been proffered for applied art.

The hermeneutic paradigms qualifying the concept of artwork have been constructed to mark hierarchical relationships between 'endogenous' and 'exogenous' origin. The distinctions between 'our' and 'other' art, or 'here' and 'there', only apparently reflect differences; in fact they have activated strong polarizations. The argument that diversity and multiplicity need to be stressed in order to counteract the danger of homogenization is certainly cogent; the claim for pluralism persuasively thrives on the awareness that universal attributes are doomed to be inadequate, unsatisfactory and ephemeral. However, while we define artistic qualities, we assess degrees of civilization. While we differentiate and distinguish aesthetic standards, we place primitive art – prehistoric as well as aboriginal – at the base of a hierarchical taxonomy and proclaim a West-centric superiority; to which, if we are European, the highest value is attached. Indeed, assigning aesthetic values is not the primary aim, when *les arts premiers* are under investigation. This expression, used to identify and localize primeval manifestations of art, has complex connotations: referring to handmade objects which were created in Africa, Asia, Oceania and the Americas, the term is enormously wide ranging and thus inevitably vague and generic. It presupposes not only the possibility, but also the convenience of adopting a large-scale, comprehensive categorization of all the artistic expressions which testify to all civilizations at their primitive stage.

What defines the specificity of Western aesthetic parameters is the strong significance attached to the notion of art as a pure entity, 'artisticity' as a value. The power engendered by artefacts is primarily mental and spiritual, not material. Art has thus been praised for its ability to inspire moral elevation, fulfil didactic aims, or express an exquisitely autotelic, self-reflexive nature. Authoritative theories have been developed to demonstrate that artistic talent thrives on the artist's unique personality, or, conversely, can be shared as an expression of collective ethos, or located in a sphere far removed from society. Art may be subject to the demands of commissioners and critics, but it will always strive to elude them. Art intertwines with the quest for truth, authenticity, pureness, refinement, which can be achieved by showing reality as it is supposed to be, or by beautifying and idealizing it. Mimetic or abstract, allegorical or symbolic, art ought to regenerate, invigorate, inspire.

Plato's notion of mimesis entails that art imitates reality which imitates the ideal world, and thus does not widen our knowledge of truth *(Republic*, X). However, it can express true values when it originates from divine inspiration, ϑεία μανία (*Ion*, 534). If poetry, painting, sculpture, music and dance are the expression of deeply inspired creativity, they enable the artist to attain a world of harmony and divine rhythms. Aristotle re-evaluated imitation by defining it not as the description of 'things occurred', but of 'things possible', which can happen 'according to the laws of verisimilitude or necessity' (*Poetics*, 1451 a-b). Such laws shed light on the world of passions and lead to catharsis (*Politics*, VIII, 7, 1341 b, 35).

Humanism produced a telluric change in the notion of creativity, shifting the focus from the manifestation of the mysterious divinity of art to the highest expression of human talent: Leonardo da Vinci celebrates the artist as creator, not imitator. In the seventeenth and eighteenth centuries, subjective dimensions conjured up by the poets' imagination and the objective reality of nature and its laws explored by science were located under different, and distinctive, jurisdictions. Romanticism reconciled two fundamental Western paradigms: art as beauty and art as truth. Idealistic aesthetics identified art with the creative activity of the Absolute, as Schelling maintained in *System of Transcendental Idealism* (1800). The ability to create and above all the gifts bestowed on genius are a continuation of God's faculties. When, during the second half of the nineteenth century, the uniqueness of artistic creation praised by the Romantics was spoiled by reproducibility, when industrial work disfigured art and architecture, artists responded by strongly advocating their autonomy or by re-appraising craftsmanship. Socialist ideologies stressed the necessity to erase the hierarchical distinction established in the Renaissance between the intellectual, inventive 'artes liberales' – architecture, painting and sculpture – able to engender original aesthetic experience, and the 'artes mechanicae', the repetitive crafts practised to produce material objects. A revaluation of the Middle Ages was instrumental in finding evidence that when the gilds of artisans were formed, project and manual labour, design and realization were profitably brought together.

Among Western beliefs a special regard has been attributed to the utopian potentiality of art, to the belief that the world can be regenerated through the subversive power of artistic creativity. However, a renewal of society triggered by a renewal of art has far greater probabilities to be fulfilled if artists have been able to benefit from a culturally rich upbringing and acquire fine technical skills. It would be naïve to ignore that such requisites are affected by social class: the capacity of artists to generate or participate in breakthroughs is more powerful, if they have had access to enabling educational opportunities. Cultural and political agency empowered by art thus presupposes the faculty of expressing art as an agent of social change. The myth of aesthetic pureness requiring elitist appreciation has certainly contributed to dimming the aura of the engaged artist. We have been taught that art is true to its nature when it remains pure and, as such, can be captured by finely cultivated minds; it is somehow diminished and downgraded when it is deprived of its almost sacred immunity by being associated with some practical function or made widely accessible.

The degree of artistic creativity and the aesthetic quality of art made and used by the people constitute the core of a controversial discourse. Indeed, enquiries into the origins of art have played a significant part in defining hierarchical standards which attribute inferior value to applied art. Evaluation criteria based on a strict distinction between pure art and material culture have been adopted to assess both Western applied art and non-Western primitive or aboriginal art: they seem to share in the production of artefacts that are employed in all human activities and are rich in symbolic and mythopoietic values appreciated by all peoples.

As anthropologists have demonstrated by studying the production of tools and artefacts in primitive communities, creativity originates from scrutiny and comprehension of nature. Knowledge of the natural world entails getting close to its phenomena, interpreting them and making use of the proximity experienced and the comprehension achieved in order to make symbolic objects usable in everyday life and in rites. Though written at an early stage of development of anthropology, when field research was conducted by means of psychological methods, according to which the researcher enacted a process of identification with the native in order to adopt the same frame of mind, James Frazer's rendition of the rituals performed by the Huichol Indians of Mexico are relevant to the topic of the origins of art, because they delve into the process through which human beings transfuse their knowledge of nature into creative activities:

> The Huichol Indians admire the beautiful markings on the backs of serpents. Hence when a Huichol woman is about to weave or embroider, her husband catches a large serpent and holds it in a cleft stick, while the woman strokes the reptile with one hand down the whole length of its back; then she passes the same hand over her forehead and eyes, that she may be able to work as beautiful patterns in the web as the markings on the back of the serpent. [1]

The act of transferring the features of the serpent's skin into a woven fabric testifies to how sources of art can be found in the production of material culture.

1 James George Frazer, 'Chapter 3. Sympathetic Magic', in *The Golden Bough: A Study in Magic and Religion* (New York: Macmillan, 1922), III ed., p. 32.

The apparently simple reproduction of a natural form involves a ritual which entails aesthetic appreciation and the capacity, enhanced by the intervention of magic, to convey the spectacle of nature into a handmade object. Far from accomplishing a merely instrumental purpose, the artefact acquires special extra-functional values, because it bears witness to how emotional response, aesthetic pleasure, magical elements and artistic creativity can interact. It would thus be restrictive to assign only to the cultural anthropologist the task of defining the features which characterise, and exceed, the materiality of the object; the contribution of anthropology of art is certainly also necessary, but still not sufficient.

A focus on origins reveals that art and material culture cannot be regarded as mutually exclusive. However, Western discourses on art are marked by a persistent demand to draw clear boundaries between the two and, more controversially, between aesthetics and ethnology. Whether artistic creativity, inventiveness, originality are inborn and can thus be expressed by all human beings, or whether they can develop only in a civilized, culturally rich environment; whether objects of use express individual creativity and elicit aesthetic fruition are still crucial, unsolved questions.

It is easy to recognize that in order to define art, human capacity to produce it must be defined. In order to attempt a definition, two thorny issues must be addressed: the first is whether humans are biologically predisposed to create art and began to do so at a very early stage of human evolution, the second is why such predisposition has remained and evolved throughout millions of years. If the idea that the creation of art initiated with *Homo sapiens* is accepted, the dichotomy between aesthetics and ethnology becomes untenable. The earliest 'when' of art is a problematic issue, but worthy of in-depth investigation, because underestimation of the origins has made it perniciously possible to disconnect the nature and representation of what is esteemed aesthetically valid or beautiful, from patterns and structures of cultural expressions, such as social organization, technology, language, mythology, religion, symbols, and rituals. The binary opposition proves particularly detrimental when the enquiry shifts from the nature to the origins of art: indeed, it proves not only fundamental, but also refreshing, to erase the distinction between objects of beauty and objects as tools, and to study why art begins with emotional drives, perceptions and mental stimuli which fuel imagination and activate creativity.

Enquiring into the problem of origins shows how necessary it is to re-assess art history as a discipline. As Whitney Davis argues in relation to the study of prehistoric art, '[...] despite offering various speculations about the meaning or function of paleolithic images, the textbooks never really address the problem of their historical origins'.[2] Conservative views are founded on the idea that, owing to the lack of archaeological evidence, the scope of discussion is restricted merely to theoretical assumptions. More innovative orientations tackle the notion of origins by underscoring the semiotic problems they pose: hermeneutic hypotheses are affected by the difficulty of decoding a system of signification which is scarcely comprehensible, because we have lost access to the conventional and symbolic meanings original makers attributed to their signs.

The challenges raised by enquiries into origins concern not only the expressive codes which indisputably still require fully effective deciphering strategies, but also the construction and circulation of cultural ideologies. Interfaced research can acquire enormous relevance if it succeeds in devising methodologies which, far from pursuing universalism, overcome the hierarchical partition between Western and non-Western art. 'What is art?', the question that humanists have incessantly raised, can metamorphose into 'when is art?', as the archaeological studies of David Lewis-Williams, Tsion Avital and Steven Mithen have demonstrated, or 'why is art?', as the anthropological and aesthetic research of Ellen Dissanayake has shown, and 'why has art originated and evolved throughout millions of years?' as the sociobiological theories of Nancy Aiken have explained.

III. Prehistory of Art

A first date for the beginning of art cannot be fixed. There is no first painting or sculpture as such. Natural pigments were used by primitive men to paint their houses, objects were made for rites and clothes and ornaments were decorated for enjoyment, but whether these activities can be defined as production of art or as acquisition of manual skills is still a matter of debate. Art versus craftsmanship is an ancient strife.

A thought-provoking contribution to the debate has been offered by David Lewis-Williams, who in *The Mind in the Cave: Consciousness and the Origins of Art* (2002)[3] explores the signs and figures painted on the walls of the caves at Lascaux and Gabillou in France between 45,000 and 10,000 years ago in order to understand how making art is significantly connected with the process through which Paleolithic *Homo sapiens* became human. By interfacing art history and cognitive archaeology, he develops two hypotheses: the production of pictures bears witness to the acquisition of a fully modern consciousness which entails

2 Whitney Davis, 'Beginning the History of Art', *The Journal of Aesthetics and Art Criticism*, 51, 3 (Summer 1993), 327–50 (p. 327).

3 (London: Thames and Hudson, 2002).

the capacity to process mental images; or, it can be defined as a manifestation of the religious system and thus employed to maintain class hierarchy. The relevance of Lewis-Williams's study lies in the intersection of archaeology, anthropology, art history, early religion, and psychology, on which he founds his method for interpreting animal and geometric imagery. In order to demonstrate that some of the wall paintings were produced by shamans, whose aim was to represent what they experienced in states of altered consciousness, Lewis-Williams discusses art and social structures in the Paleolithic Age by availing himself not only of the tools of anthropology of art, but also of neurology.

Methods of dating the beginnings of pictorial representation supported by the study of cognitive structures can change our view of the nature of art. The specific cognitive capacity which gave rise to prehistoric art could be of very early origin, probably earlier than the human ability to make tools. The fact that very similar representational systems appeared in all hunting societies could be regarded as evidence supporting the theory that such cognitive capacity was common to all hunters everywhere. Tsion Avital has defined it as the ability to read and interpret footprints. [4]

Art, considered as an ability which pertains to human beings and makes itself manifest when an object is created, infused with distinctive qualities, aesthetic or practical, and enriched with symbolic values, involves a deep capacity for connection, enacted through transmission, reception and response. Art is thus a uniquely powerful form of establishing connections between people. By exploring the idea that art is the most complex and enticing medium of human communication, in *The Prehistory of the Mind: A Search for the Origins of Art, Religion, and Science* (1996) Steven Mithen argues that artefacts or images created to communicate complex and symbolic meanings are a product of the cognitive fluidity which marks the difference between the 'Modern' and the 'Early' Human Mind. The ability to interpret 'natural symbols' such as hoofprints ('natural history intelligence'), intentional communication ('social intelligence'), and the ability to produce artefacts from mental templates, e.g. a stone handaxe ('technical intelligence'), are the three cognitive capacities critical to making art, which were present but still separated in the mind of Neanderthal man, defined in this way as Early Human. These capacities came to interact and enhance each other in the Modern mind of *Homo sapiens sapiens.* [5] The growing fluidity of cognitive processes, by fostering the development of relationships

4 Tsion Avital, 'Footprints Literacy: The Origins of Art and Prelude to Science', *Symmetry: Art and Science* (The Quarterly of the International Society for the Interdisciplinary Study of Symmetry), 9, 1 (1998), 3–46.

5 Steven Mithen, *The Prehistory of the Mind: A Search for the Origins of Art, Religion, and Science* (London: Thames & Hudson, 1996), pp. 151–63.

between specific forms of intelligence, gave rise to the production of objects exhibiting instrumental as well as ornamental characteristics.

The most recent theories of the origins of artistic creativity focusing on its prehistory and founded on interfaced methods also call for a reassessment of the notion that art began with the production of representational and figurative signs, like the ones in the Lascaux cave paintings. Objects with simple geometrical markings have not been regarded as early art objects, because they are non-representational; rather, and far too restrictively, they have been considered as tools, or manifestations of productive activity, or incongruous items. However, the hypothesis according to which the origins of art making are to be found in figurative signs dating to a period as recent as between 45,000 and 30,000 years ago ignores the contributions of men who lived in much earlier periods and the importance of early abstract art. [6] It is thus crucial to re-discuss the assumptions according to which the beginning of art coincides with the beginning of figurative representation.

IV. Origins and Evolution of Art

Evolutionary views of art were first enunciated by Herbert Spencer in *The Principles of Psychology* (1882), where he claimed that the notion of evolution could explain the origin and persistence of artistic creation and experience. While evolving, human beings have increased their inborn capacity to accumulate more energy than is necessary for survival. The unspent surplus is stored and available to be channelled into artistic creativity. While maintaining that the permanence and development of art is due to the release of energy progressively accumulated by human beings, this theory does not provide an explanation for the varieties of artistic expressions. [7]

During the first half of the twentieth century the epistemological validity attributed to taxonomies encouraged numerous philosophers to articulate sophisticated systems of the arts. *System der Aesthetik* by Johannes Volkelt (1905), *Système des Beaux-Arts* (1920) and *Propos sur l'Esthétique* (1923) by Alain, *Aesthetik und Allgemeine Kunstwissenschaft* (1923) by Max Dessoir, *The Principles of Art* (1938) by R. G. Collingwood, *The Arts and the Art of Criticism* (1940) by Theodore Meyer Greene, *La Correspondence des Arts* (1947) by Etienne Souriau, *The Arts and their Interrelations* (1949) by Thomas Munro, *The Modern*

6 See Bill Dolson, *Notes on The Origins of Nonrepresentational Art*, <http://www.billdolson.-com/text/origins.htm> [accessed 15 August 2009].

7 See F. Graeme Chalmers, 'The Study of Art in a Cultural Context', *The Journal of Aesthetics and Art Criticism*, 32, 2 (Winter, 1973), 249–56 (p. 249).

System of the Arts (1951–1952) by Paul Oscar Kristeller, *Traité d'Esthétique* by Raymond Bayer (1956) enunciated all-encompassing models of classification, which re-examined historical definitions of the arts in bipolar terms such as 'major and minor', 'pure and applied', 'spatial and temporal'. These syntheses situated the arts on a purely abstract, metaphysical plane and throve on the appropriation of notions introduced by *Geistesgeschichte*, the history of culture regarded as the history of the essence of humankind throughout the centuries. In his seminal, influential essay *Symbiose der Künste* (1937) Kurt Wais declared that the artist constantly strives to achieve a total work of art, but human faculties, even the more versatile, only suffice for the creation of fragments of beauty.

> The primary and most difficult act of creation is the decision to shatter the original symbiosis of all arts, the (often unconscious) choice to paint (malen, make a mark, fix), to write (dichten, dictare, set down), or to compose (put together). Some never find the strength to make this decision, but, like the young Hoffmann, consume themselves with crippling uncertainty, wondering whether they are meant to be painters or composers. A certain memory of the original unity affects nearly all of them. These enduring ultimate bonds among the arts betray themselves in numerous cases of multiple talent in artists [...].[8]

The creative process generates strife between the primeval unity of the arts and the individuality of the artist who, though doomed to fail, yearns to recreate an imagined totality. The appeal of Wais' arguments lies in the assumption that there exists an ideal, purely aesthetic realm, where arts live in full integrity and perfection. Such abstraction stems from Kant's belief that art exists in order to exist and not to be useful. The idea that our fine arts are non-utilitarian, still deeply rooted in philosophy, needs to be thoroughly questioned, precisely because it has fuelled the bias according to which 'our' art is above 'other' arts.

The evolutionary perspective has proved valuable for deconstructing discriminatory paradigms. According to Ellen Dissanayake, what distinguishes humans are certain characteristics which, though present in their animal ancestors at an elementary stage, show a peculiarly human degree of development and interrelatedness. The virtually unlimited and constantly evolving capacity to interweave each characteristic in order to activate new attitudes, interactions,

8 Kurt Wais, 'The Symbiosis of the Arts', Address to the Second International Congress of Modern Literary History (Amsterdam: September, 1935), trans. by Gregg A. Richardson, *Yearbook of Comparative and General Literature*, 31 (1982), 78–95 (p. 85). The paper was originally published in German as *Symbiose der Künste, Forschungsgrundlagen zur Wechselsberührung zwischen Dichtung, Bild- und Tonkunst*, Schriften und Vorträge der Württembergischen Gesellschaft der Wissenschaften: Geisteswissenschaftliche Abteilung, 1 (Stuttgart: Kohlhammer, 1936), and then as 'Vom Gleichlauf der Künste', *Bulletin of the International Committee of the Historical Sciences*, 37 (1937), 295–304.

and behaviours allows humans to perform all activities, among which are the arts:

I see these integral components of all human behaviour as filaments or skeins that can be interlaced in different combinations to make cables or braids that then combine into more complex behaviours, such as those that make up the arts (e.g. songs, poems, stories, decorated objects, and so forth). What I mean will be illustrated after artificially isolating and describing each general behaviour or predisposition. They can be said to be innate powers and tendencies—products of natural selection during the aeons of hominid evolution. They require the setting of a particular culture in order that their potentiality can fill specific roles that will be performed in an integrated, specific way. They are intrinsic and inextricable ingredients of human nature.
Each component has contributed in its way to art, whose versatile garment shows, under the embroidery and ornaments of its many uses, the multicoloured threads and filaments of its primordial substance. The elements of art are human nature's fundamental elements, though often, in modern Western art, they appear in non-functional manifestations. [9]

Dissanayake wants to deconstruct the belief that artistic creativity is exquisitely gratuitous by demonstrating that it fulfils special functions, which humans, and only humans, can appreciate. Because it is a predisposition which belongs to humans, and to no other species, it qualifies their very nature and stands out as one of their distinctive manifestations. Art defines them as being human and in so doing, it shows how, far from being an extension of humanity, it constitutes their biology, that is the characteristics and behaviours which are specific and different from those of other living organisms. Dissanayake identifies stimuli which produce reactions at a very early stage of human life. The formation of aesthetic experience is characterised by the widely different degrees of intensity and interaction generated by the stimuli, but the ability to respond to them is innate. While pointing out that both the physiology and psychology of the subject stimulated are affected deeply and in different ways, she also emphasises the very simple and basic nature of the stimulation.

Just as the roots of the arts were found to derive from general intrinsic behavioural endowments or 'filaments', the primordial sources of aesthetic experience can be said to inhere in reflexive general responses to biologically significant stimuli. [...] the infant human species first displayed its nascent aesthetic sensibility in reacting to pattern, to repetition, to novelty, variety, rhythmic sequence, intensity, balance,

9 Ellen Dissanayake, *What is Art For?* (Seattle and London: University of Washington Press, 1988), p. 111.

and other effects closely associated with the physiological and psychological processes common to all living creatures. [...]
While the occurrence in art of such elements would suggest that at least some of our responses are if not automatic then preordained or facilitated by our biology, there is certainly more to aesthetic experience than response to biologically relevant stimuli.[10]

Dissanayake, aware that positing the biological bases of artistic activity engenders the hazardous issue of determinism, specifies that, while art performs notable functions, its functionality also varies according to individual traits and to the cultural milieu in which it can be developed and expressed.

In *Homo Aestheticus: Where Art Comes From and Why* (1995) Dissanayake firmly rejects the claim for non-utilitarian art by arguing that, since it is shared and performed in every epoch and culture, and persists in all human societies, it must have some utility. Ethology provides conceptual paradigms and critical tools which can be fruitfully adopted in the study of the origins and evolution of the arts. A universal inherited propensity to make some objects and activities special is what presides over human capacity to produce and express art.

The ethological point of view entails demonstrating that there are biological components involved in the production of art, namely adaptivity, that is permanence through evolution, and universality. Why art has evolved and is a universal distinctive trait in the human species can be explained by arguing that it establishes a network of fundamental structural relations such as sharing, mutuality, and a sense of belonging. A theory of art founded on bio-behavioural premises aims at overcoming assessments exclusively, and reductively, based on aesthetic categories. 'What is art?' does remain a relevant question, among other crucial ones, such as 'what is art for?', 'what does art do to, and for, human beings?'. Certainly the notion of universality, adopted to qualify the biological origin of artistic behaviour, must be disentangled from possible associations with homogeneity: the contention that artisticity is universal encloses the awareness that human beings are diverse. The attempt to prove that the universality of art as a behaviour intrinsic to and shared by human beings coexists with the diversity of its production and fruition demands that ethological theories are supported by a perspective able to reach a plurality of cultures. The necessity for a broad view is certainly acknowledged by Dissanayake, whose conceptions have emerged out of her field research in Skri Lanka, Nigeria, India, Madagascar, and Papua New Guinea.

A few years later Nancy E. Aiken published *The Biological Origins of Art* (1998), which incorporates Dissanayake's work. Aesthetics, psychology, neu-

10 Ibid., p. 150.

roscience, and sociobiology are the disciplines Aiken explores to demonstrate that art entails psychic responses which are generated also by biological factors. Further stress is laid on the idea that, by studying behaviour from the viewpoint of evolution, ethology provides the conceptual framework for exploring why, after having been moulded by evolutionary forces, the human behaviour called art making has persisted and fulfilled specific purposes. The invitation to view the production of art as behaviourists view animal activities may sound provocative, but its value resides in the plea for a radical reassessment of traditional critical parameters.

Although her theories are based on the assumption that human behaviour is shaped by evolutionary processes, she dispels the ghostly presence of determinism:

> I try to demonstrate that human behaviour is shaped, moulded, or constrained by our biological makeup, but is not set or predetermined in the sense that our behaviour is made automatic by our nervous system. [...] Also, I [...] describe a particular kind of reflexive behaviour, the defense reaction, in order to demonstrate that (1) the defense reaction is part of the human behaviour, (2) the defense reaction is reflexive in nature, (3) emotion is part of the defense reaction, and (4) some of the same natural stimuli which trigger defense reactions in human beings are used in art to trigger emotional responses which are part of the defense reaction and which are evaluated by us as aesthetic response. I then discuss various stimuli which probably are used in art to trigger the emotional portion of the defense reaction, and some stimuli which may trigger other emotional responses.[11]

Aiken is not contending that aesthetic experience is the outcome of reflexive responses, but rather that a degree of reflexivity, such as the reaction to certain lines, shapes, colours or sounds, is involved and activates a synergy with other responses. The impact of the hypothesis formulated by Dissanayake and expanded by Aiken resides in the crucial cultural implications attached to their biologically based views of the origin of art. The biological built-in predisposition to respond involves the notion that aesthetic appreciation is not only a shared human trait, but also, and more remarkably, that it presents universal qualities. Commonality and universality would thus explain why art produced in different historical and cultural contexts can be appreciated: Aiken stresses that the reflexive behaviours we display with regard to certain stimuli are not different from their very first manifestations, which occurred many thousands, and possibly millions, of years ago.

11 Nancy E. Aiken, '1. Big Questions', in *The Biological Origins of Art* (Westport: Praeger, 1998), p. 7.

In order to avoid the risk that the pathways opened up by sociobiology reveal obscure zones, any possible association with extremist neo-Darwinian stances must be dismissed. The focus laid on the biological origins of artistic production and appreciation is aimed at showing the inadequacy of the hierarchical assumption that art can be credited with better or worse qualities according to degrees of civilization or primitiveness attributed to the culture within which it is produced. Nonetheless, the contention, deriving from the more fundamentalist districts of sociobiology, that genes play a central role in human behaviour, and thus variation in attitudes and responses can be explained in terms of peoples' biology rather than of the social environment they live in, sounds unacceptable. Instead, theorising art as the product of human behaviour, triggered by specific biological predispositions and nourished by peculiar environmental factors, opens up perspectives which can overcome the limitations and biases inherent in theories of immutable, autotelic artisticity.

V. Re-Directing

Our culture has shown a persistent inclination to classify the arts in terms of 'high' and 'low': such distinction has conveniently accommodated popular art, a typology which has aroused constant apprehension, above all since the process of commodification following the Industrial Revolution has created mass art. Popular art encompasses artefacts characterised by a genesis and a destination which does not require the deployment of fine aesthetic parameters.

The highest speculative value granted to the study of the fine arts is finely complemented by the cultural interest attributed to the observation of handmade objects. The bipolarity deserves closer investigation. Indeed, the characteristics we attach to Western 'popular' art exhibit remarkable similarities with the features we note in the art of 'non-Western', 'primitive' cultures. 'Popular' is the label conventionally assigned to woodcarving, embroidery, painted glass, metal working and every other form of craft, while the classification of civilizations in terms of 'non-Western' and 'primitive' is utilized, not less conventionally, to encompass Africa, Asia, Oceania and Americas – all of which the Musée du Quai Branly in Paris claims to represent – and to include artistic expressions not properly definable as 'pure' art. In fact, those artefacts may be made to be beautiful, but are certainly, and possibly primarily, intended for specific uses. Our paradigm of art hides a dangerously West-centric ambivalence: Art, to be artistic and aesthetically valid, must be created spontaneously, for the sake of Beauty; it ought not to be used but exhibited, contemplated, transformed into an aesthetic discourse; or, it can be inscribed in a wider social design, or enforce political ideologies or regime propaganda, but only if it abdicates its pureness.

A paradigm shift becomes possible, and even inevitable, by exploring other approaches to the origins of art, which I propose to define as humanistic-scientific in order to indicate that they commingle. United, they go beyond the paradigms traditionally employed by philosophy and history of art.

It would be unfair to state that our philosophers have been persistently unaware of the necessity to overcome judgements about aesthetic quality based on the pureness or instrumentality of art. In his essay on 'Der Ursprung des Kunstwerkes' ('The Origin of the Work of Art', 1935 – 1936) [12] Martin Heidegger argues that when we examine something we provisionally classify as an object, the fundamental question is about the relationship it establishes with 'World', to which the essence of being pertains, and with 'Earth', which hosts objective existence. Though essentially different, they are never separate. Works of art, possessing the special quality of being related to both World and Earth, express the conflict between them. As Giovanni Bottiroli points out, powerful dynamics between World and Earth, or, according to his definition, between articulation and density, can originate from an object created to serve instrumental purposes and usually classified as belonging to the crafts, or can be found missing in an object created to be pure art but lacking internal conflict. [13]

The essence of the work of art is its origin, not in a historical or chronological, but ontological sense: it is its *Ur-sprung*, its capacity to leap beyond its historical and cultural context. The true artworks are recognisable owing to their '*Ur-sprung* quality', namely their mysterious trans-epochal nature, their macro-temporality, in Bachtin's terms. Thus the universality of the masterpieces, far from being an anti-historical immobility, a rigid, un-violated, crystallised and owned permanence, is a dynamism which pierces empirical time.

Ontological and ontic are different, as the former pertains to the essence, the latter to the manifestation; however, a relationship of co-belonging can be established. Ontology does not aim at eluding or overcoming ontic concreteness, which is the research field of linguists, historians and anthropologists. If it did so, it would cause a split, which Heidegger certainly rejects. His fascination with 'essence' could raise the concern that he ends up constructing an aristocratic aesthetics of purity, but what he achieves is exactly the opposite. His approach to the origin of art proves viable in so far as it disregards hierarchical parameters and indicates that the distinction between pure and applied arts is devoid of

12 Martin Heidegger, *Der Ursprung des Kunstwerkes* (Ditzingen: Reclam Universal-Bibliothek, 1986); *Holzwege* (Frankfurt: Klostermann, 2003).

13 Giovanni Bottiroli, *Teoria della letteratura* (Torino: Einaudi, 2006), in particular chapter VI, *8. L'opera d'arte. Ancora la 'e' di 'Essere e tempo' – Arte e verità – Il conflitto fra Mondo e Terra (articolazione e densità)*, pp. 375 – 87.

meaning. Nonetheless, ontology explores questions and adopts conceptual paradigms which are different from the pursuits and methods of natural and social sciences.[14]

In recent years the assumptions on which Western views of art are based have undergone close scrutiny by non-Western critics. In the 2006 essay entitled *Historical Constructions of Arab Art and Visual Culture* Maymanah Farhat, who promotes knowledge of Middle Eastern culture and the work of Arab Women Artists, provides a very lucid synthesis of the imperialist and colonialist ideologies which have marked our understandings of other art. She underscores that, instead of facilitating the circulation and exchange of knowledge about different modes and contexts of artistic production, art history has erected barriers and evaluated degrees of superiority and inferiority, by strictly asserting what can be included in the definition of art, and what should be excluded.

The effectiveness of her critical discourse, which gains rhetorical impetus through long, powerful sentences and tightly interlocked arguments, sounds like a punctual counter-response to the forcefulness of the art ideologies imposed by our culture. By drawing a geo-political map of the continents in the nineteenth century, she points out how research conducted in the fields of archaeology and anthropology, and evaluation parameters subsequently utilized by art history, were intricately connected with European imperialism and colonialism:

> Censorship has impacted understandings of non-Western (African, Asian, Middle Eastern, Pacific Islander, Latino and Native American) art and visual culture for centuries. Art historical classification and documentation of non-Western art and visual culture has been influenced by archaeological studies conducted during the colonial and imperial endeavors of nineteenth-century European governments. Masking intentions of economic exploitation and political dominance, colonial and imperial activities were justified by propagations of racial, cultural and theological superiority.
>
> Reflecting such notions, archaeological studies of the time formulated a denigrated categorization of indigenous art and culture. Evidence of early indigenous artistic production was classified as belonging to the expansive, all encompassing category of 'Ancient Art' instead of attributing it to that of the specific non-Western culture in which it was found. Such classification, coupled with the emergence of anthropology in the nineteenth century [...], labelled indigenous art and visual culture as 'primitive' or 'naiveté.' With a growing increase in ethnographies and travel between the West and its colonies, the demand for the material consumption of 'exotic' culture led to the exportation of 'objects' made by the 'other' for Western consumption, reducing indigenous art forms to nothing more than commodities. During this time, art history materialized as an academic discipline, using ar-

14 I wish to thank Giovanni Bottiroli for the generosity with which he has shared his insightful critique of Heidegger's views about the origin of the work of art.

chaeological and anthropological 'findings' and categories as a basis for the historical, conceptual and aesthetic analysis of non-Western visual culture. Deeply rooted in colonial and imperial sociopolitical traditions, art historical discourse has been plagued by institutionalized cultural hierarchy that retains non-Western art as substandard to its Western equivalent. [15]

New hermeneutic categories are required to deconstruct the assumption that non-Western art variously labelled as prehistoric, primitive, early, ancient, indigenous, or aboriginal is not pure art. Hierarchical differentiations prove inadequate and above all limited, if one thinks how productive art criticism can be by focusing on pluri- rather than univocality.

Because the topic of art origins is eventually receiving attention, critical discourses are proliferating. How alluring they can be is particularly evident in the Parisian Musée du Quai Branly: outside, a magnificent encasement of glass and vegetation – 'le mur végétal' ('the living wall') – frames the administrative offices [Figure 3 and Figure 4]; inside, visitors travel across cultures while enjoying intense visual stimuli aroused by a sophisticated, high-tech atmosphere. That the lure can be easily detected, and rejected, is evident in the article published by Michael Kimmelman on *The New York Times* in 2006:

The most ambitious museum to open in Paris in 20 years, dedicated to non-European cultures, Quai Branly provoked a ruckus from the instant President Jacques Chirac came up with the idea for it more than a decade ago. It was his monument to French multiculturalism and, perhaps, to himself. Two beloved Paris institutions had to be dismantled, the Musée des Arts Africains et Océaniens and the ethnographic department of the Musée de l'Homme, France's sublime natural history museum. Anthropologists, not to mention more than a few people who loved going to those museums, were furious. The familiar aesthetics-versus-ethnology question came up: Will religious, ceremonial and practical objects, never intended as art in the modern, Western sense, be showcased like baubles, with no context? [16]

Non-Western artefacts, exposed to views which purport to be aesthetic but in fact, and inevitably, bear the mark of cultural ideologies advocating Western supremacy, end up encapsulated within artificial, strained critiques. Adjusted in

15 Maymanah Farhat, 'Historical Constructions of Arab Art and Visual Culture', excerpted from the exhibition catalogue *Three Arab Painters in New York: Samia Halaby, Sumayyah Samaha and Athir Shayota*, copyrighted to Maymanah Farhat and Al Jisser Group, May 2006, <http://www.arteeast.org/artenews/artenews-articles2006/historical-constr7-06.html> [accessed 15 August 2009].

16 Michael Kimmelman, 'ART; A Heart of Darkness in the City of Light', *Times*, 2 July 2006, <http://www.nytimes.com/2006/07/02/arts/design/02kimm.html> [accessed 15 August 2009].

Figure 3: Le mur végétal, Musée du quai Branly, Paris, 2006. Architect: Jean Nouvel; Green Wall Designer & Installation: Patrick Blanc.
Figure 4: Le mur végétal, Musée du quai Branly, Paris, 2006, detail.

apparently spacious, actually meticulously bordered settings, they become misshapen, abnormal and inaccessible. The museum, which takes up the vertiginous challenge of presenting collections of objects belonging to the civilisations of Africa, Asia, Oceania and the Americas, becomes a hyperbolic site, replete with signs and symbols, but at the same time built to locate, arrange, enshrine. The ultimate paradox is that, though finely shaped to trigger invigorating cultural experiences, the museum where 'the cultures dialogue' becomes a vacuum which weakens and rarefies the power of the objects exhibited.

Humanistic-scientific views on non-Western art have the potential to respond by complementing, rather than replacing, historically and philosophically oriented conceptions. Indeed, the notion of interface presupposes the activation of osmotic processes which can neutralise biased, hierarchical preconceptions about fine and lesser, pure and applied, us and them.

Lewis-Williams, Mithen, Avital, Dissanayake and Aiken have demonstrated how trans-disciplinary enquiries into the origins of art can thrive on methods which, while maintaining the efficacy of critical modes based on history and philosophy, encompass archaeology, anthropology, cognitive sciences and biology. The ample scope adopted by these critics for observing cultural specificities and comparing attitudes toward art is a common feature resulting from a shared alertness to the risk of hazardous generalizations. Because origins have been eluded by humanities scholars, journeying through these neglected areas stimulates the construction of a new cartography of the arts. Interfaced methodologies are worth developing because they can offer a substantial contribution towards an understanding of art as an intrinsic, and shared, rather than a privileged, almost exclusive, human expression.

Recent studies in archaeology have demonstrated that the origins of art must be related to the gradual development of cognitive processes which enabled primitive humans to connect various forms of intelligence to produce objects with ornamental and instrumental characteristics.

Unencumbered by biological determinism, which would cause a relapse into early twentieth-century social Darwinism and eugenics, sociobiology has reconsidered the history of evolution in order to prove that art originates from specific emotional stimuli and, because it shows characteristics of adaptivity, can be regarded as an expression of biological survival.

Recent studies in ethology have led us to acknowledge that art is a universal behaviour aimed at establishing relationships like correspondence, sharing, exchange, sympathy; within this prospect, our aesthetic categories do not suffice, because they encapsulate art within a reductive framework of values.

Particularly teasing questions such as 'what is art?' or 'what is the true essence of art?' can be overcome by raising other questions such as 'what cultural changes can be produced by regarding art within an evolutionary and ethological perspective?'. It is a commitment which cannot neglect, of course, the dangers of a universally homogeneous view of art, but must focus precisely on diversity. It is an enterprise to be undertaken without naiveté and with the awareness that the art of Papua New Guinea remains difficult for us to decipher, but nonetheless is worthwhile serious exploration not only *per se*, but also because it offers the possibility of deconstructing the dichotomy between aesthetic and ethnological.

Traditional aesthetics have posited the existence of a self-enclosed, self-sufficient, self-reflexive realm of art and defined universal, timeless criteria of artistic value. Humanistic-scientific approaches to the origins of art explore how the mind works and which cognitive processes trigger appreciation of beauty.

Shifting the focus from 'origins of' to 'origins in' art allows for an appreciation of how artists themselves respond to the quest for beginnings. Twentieth-century figurative artists express meta-artistic, self-reflexive views in representations of origins which appear beguilingly simple but enclose complex interpretations. In *Beginning of the World* (1920) [Figure 5] by Constantin Brancusi the essentiality of the forms contrasts with the complex nature of the subject. The origins he concocts highlight the contrast between the curvilinear shape of the ovoid object and the sharp square and rectangular lines of the pedestal. While showing how powerfully ancestral times can be evoked through visualization of an emblematic primeval shape, his rendition is also an appropriation of other representations; it is a powerful expression of the Modernists' re-conceptualisation of the primitive. Re-located in a modern context, a primordial form retains its appearance but is charged with new meanings. The extraordinary power by which an apparently simple object can uncover multiple facets is exhibited by Anish Kapoor in *Turning the World Inside Out* (1995) [Figure 6]: the apple is ordinary and obvious, but also endowed with novel and exceptional qualities. By working on the volume and material of a quintessential primeval thing, he creates a centre which discloses multiple refractions and re-figurations of reality. *Empezando a echar raíces al otro lado (Beginning to grow roots on the other side*, 2002) [Figure 7] by José Bedia can be defined as a quest for modes of interfacing which shun hierarchical relationships while interacting with diversity. Will the mutant, though equipped with extra long limbs, succeed in bridging the gap, or will he be run over by the oncoming vehicle? Positive answers may sound facile, because the erosion of deeply rooted conceptions about the source and essence of art shakes the core of our culture. Yet, because answers need to be envisaged, the humanities and the sciences can benefit from testing the knowledge each of them has acquired and then sharing new re-sources.

Searching for the sources of creativity entails identifying the stimuli, perceptions, and mental elaborations which give rise to the artistic imaginary. As Aiken points out, 'Aestheticians in cultures as diverse as the Japanese, the Chinese, the Indian, and the Western all seem to be describing the emotional effect of art'. [17] Given this acknowledgement, it is clear that new answers as to questions about how art affects human beings will be formulated by humanists and scientists. And, rather than or: not a belligerent confrontation, but a wider aegis.

17 Aiken, p. 23.

Figure 5: Constantin Brancusi, *Beginning of the World*, c. 1920, marble, nickel, silver, and stone (76.2 x 50.8 x 50.8 cm), Dallas Museum of Art, Foundation for the Arts Collection, gift of Mr. and Mrs. James H. Clark. Photograph © 2010 Dallas Museum of Art.

Figure 6: Anish Kapoor, *Turning the World Inside Out*, 1995, cast stainless steel (3/3, 152.4 x 177.8 cm), Santa Barbara Museum of Art, Museum purchase, 20th Century Acquisition Fund and funds provided by Eli and Leatrice Luria and the Luria Foundation, Jon and Lillian B. Lovelace Jr., Smith Richardson and the Grace Jones Richardson Trust, and the SBMA Visionaries. Photograph © 2010 Santa Barbara Museum of Art.

Figure 7: José Bedia, *Empezando a echar raíces al otro lado (Beginning to grow roots on the other side)*, 2002, lithograph printed in brown, grey, black and white from four aluminium plates on cream Rives BFK paper (76.2 x 112.4 cm), Boston, Museum of Fine Arts. Lee M. Friedman Fund, 2003.625. Photograph © 2010 Museum of Fine Arts, Boston.

Andrea Grignolio / Vita Fortunati

Memory as the Origin of Biological and Cultural Identity in the Work of Samuel Butler [1]

Recent researches have thrown absolutely no light upon the origin of life – upon the initial force which introduced a sense of identity and a deliberate faculty into the world.

Samuel Butler, *Lucubratio Ebria* (1865)

Let us at once go back to the primeval condition of the race.

Samuel Butler, *Darwin Among the Machines* (1863)

I. Introduction

The issue of the memory of origins is a sort of *file rouge* in the works of Samuel Butler (1835 – 1902). From *Erewhon* (1872) to the posthumous *The Way of All Flesh* (1903), passing through anti-Darwinian books *Life and Habit* (1878), *Evolution, Old and New* (1879), *Unconscious Memory* (1880), and *Luck, or Cunning?* (1887), Butler's main concern is the dependence of personal identity both on biological instincts (inheritance) and social as well as familial traditions (heritage). According to Butler, personal identity, as shaped by biological and cultural memories, can only act under direct influence of past experiences accumulated by its ancestors. Each individual inherits unconscious memory and transmits it across generations thanks to a typical Lamarckian hereditary mechanism able to transform organismal willingness into habits, habits into instincts, and instincts into memories. Inspired by Ewald Hering's previous model, Butler called it the organic memory theory.

As a Victorian, Butler was not the sole author insisting on the myth of individuality as a means for social progress; what makes him a central figure in the evolutionary debate on selfhood was his growing up in the same milieu of

1 Sections I, V, VI and VII have been written by Andrea Grignolio, sections II, III, and IV by Vita Fortunati.

Charles Darwin or, as it were, under his shadow. Indeed, he was first a great admirer, a family friend and a frequent visitor at Down House, then became a strong opponent of his natural selection theory and, eventually, of Darwin himself. The notoriety brought about by his rejection of Darwinian consensus and the theory according to which the power of individuality can alter the course of evolution have fostered a rediscovery of his thought as a third force in Victorian intellectual life, as opposed both to scientific naturalism and to religious orthodoxy.

Two reasons were at the root of Butler's firm opposition to Darwin. Firstly, by introducing the principle of natural selection, Darwin eliminated purposefulness, i. e. teleology, from the evolutionary process: the perspective of a materialistic, chance-driven Nature left Butler no room for his theory, according to which organismal striving (willingness, in the case of humans) is the major cause of biological and social progress. Secondly, Butler accused Darwin of having 'violated the standards of literary courtesy' by circulating his forthcoming anti-Darwinian book within the evolutionist intellectual milieu so as to soften its impact on the audience.

Darwin's refusal to offer either an excuse or an explanation in response to the accusatory letters that Butler sent to local newspapers generated an even fiercer hostility. Butler launched new attacks based on the belief that Darwin, again accused of dishonesty, failed to attribute the principle of evolution to its actual founding fathers: Georges-Louis Buffon, Jean-Baptiste Lamarck, and Erasmus Darwin. The insistence on the influence of Erasmus Darwin over his grandson was for Butler instrumental both in supporting his belief in the self-willed acquisition of traits (an idea strongly championed by Erasmus Darwin who admittedly borrowed it from Lamarck), and in confirming his own organic memory theory about the influence of cultural and familial memories on offspring. The legacy of Erasmus' biological thought over his grandson Charles was an emblematic example. As in the case of the dependence of individual identity on its origins, Butler employed the organic memory theory to trace the concept of evolution back to its original forefathers.

The concept of organic memory, vaguely sketched in *Erewhon* and then clearly developed in his four anti-Darwinian books, enabled him to challenge the dominant familial, social, and scientific orthodoxy of his time. By drawing ambiguous and paradoxical analogies, which can be regarded as a distinctive rhetorical feature of his writing, Butler skilfully blurred the boundaries between biological and cultural memories, merging personal history and familial heritage, individual instincts and ancestral memory, national identity and racial origins.

II. The Cultural Background

Butler did not enjoy the fame he desired: [2] the only successful novel was *Erewhon* (1872), which Charles Darwin appreciated for its pungent satire. [3] One of the reasons for his missed fame is partly due to his eccentric character and constant disagreements not only with his family, but also with Darwin, against whom he waged a seemingly endless controversy. Butler was an awkward, restless character, endowed with a somewhat bizarre genius. He had an eclectic mind, was interested in countless fields, from painting to photography, from Greek literature to the themes of memory and evolution, to which he dedicated many years of his life.

While reading Darwin's *Origin of Species* (1859) during his stay in New Zealand, Butler elaborated his theory on unconscious memory and started an imaginative process which led to the writing of *Erewhon*. In this sense he is an excellent example of the fruitful late-nineteenth-century exchange between literature and science, not only as regards themes, but also descriptive strategies and tropes, in particular metaphor. As Laura Otis states: 'Science was not perceived as being written in a 'foreign language' – a common complaint of twenty-first-century readers. As a growing system of knowledge expressed in familiar words, science was in effect a variety of literature. Scientists knew that like literary writers, they relied heavily on imagination'. [4]

Another reason for Butler's lack of popularity is to be found in his use of paradox. It is not actually easy for the reader to follow his tortuous reasoning and particularly to manage the oppositional strategies of his discursive rhetoric. Peter Raby contrasts Darwin's methodical and exact style with Butler's, characterised by speculations and digressions: 'While Darwin was the supreme example of a methodological scientist, Butler operated quite differently as a speculative philosopher or theologian'. [5]

Most of the times it is difficult to understand if Butler is making fun of the reader, or whether he is speaking seriously. For example, in the initial pages of his first work on evolution, *Life and Habit* (1877), Butler wears the mask of someone interested in science without knowing much about it, a sort of man in the street who enjoys speculating on phenomena:

2 Lee Elbert Holt, 'Samuel Butler and His Victorian Critics', *ELH*, 8, 2 (June 1941), 146–59.
3 Samuel Butler, *Erewhon, or Over the Range* (London: Penguin, 1872 [1985]).
4 Laura Otis, 'Introduction' to *Literature and Science in the Nineteenth Century. An Anthology*, ed. by Laura Otis (Oxford: Oxford University Press, 2002), pp. xvii, xxi. See also Gillian Beer, *Open Fields: Science in Cultural Encounters* (New York: Oxford University Press, 1966).
5 Peter Raby, *Samuel Butler. A Biography* (London: The Hogarth Press, 1991), p. 161.

In the outset, however, I would wish most distinctly to disclaim for these pages the smallest pretension to scientific value, originality, or even to accuracy of more than a very rough and ready kind [...]. I have no wish to instruct, and not much to be instructed; my aim is simply to entertain and interest the numerous class of people who, like myself, know nothing of science, but who enjoy speculating and reflecting (not too deeply) upon the phenomena around them. [...] It is plain, therefore, that my book cannot be intended for the perusal of scientific people; it is intended for the general public only, with whom I believe myself to be in harmony, as knowing neither much more less than they do. [6]

Butler felt offended when reviewers of his books accused him of being an amateur and consequently did not take him too seriously. This is one of his many contradictions: an intricate personality, torn between the desire of stating his ideas against his father, who wanted him to enter the Church, and the impossibility of openly declaring his homosexuality in a society ruled by rigid conventions. These autobiographical elements allow for understanding both the several disguises he wore and the inconsistent and enigmatic aspects of his work. In *Life and Habit* he did not uphold a monolithic concept of identity, [7] but a multifaceted one: it is a concept in many ways already typical of the twentieth century, as recent studies on Butler's importance in the development of modern evolutionary psychology have shown: [8]

We regard our personality as a simple definite whole; as a plain, palpable, individual thing, which can be seen going about the streets or sitting indoors at home; as something which lasts us our lifetime, and about the confines of which no doubt can exist in the minds of reasonable people. But in truth this 'we', which looks so simple and definite, is a nebulous and indefinable aggregation of many component parts which war not a little among themselves, our perception of our existence at all being perhaps due to this very clash of warfare, as our sense of sound and light is due to the jarring of vibrations. [9]

Butler's four books on evolution are, like his dystopian novel *Erewhon*, hybrid works, in which elements typical of the novel, the essay and autobiography commingle. In *Erewhon* he widely uses irony and parody in presenting the habits and customs of the utopian settings which should appear as a sort of Victorian

6 Samuel Butler, *Life and Habit* (New York: AMS Press, 1968), pp. 1 – 2.

7 For the crisis of the concept of individualism see *Embodied Selves: An Anthology of Psychological Texts 1830 – 1890*, ed. by Sally Shuttleworth and Jenny Bourne Taylor (Oxford: Clarendon Press, 1998), p. xiv.

8 Ruth Parkin-Gounelas, 'Mind Matters: Butler and Late Nineteenth-Century Psychology', in *Samuel Butler. Victorian against the Grain: A Critical Overview*, ed. by James G. Paradis (Toronto – London: University of Toronto Press, 2007), pp. 195 – 219.

9 *Life and Habit*, p. 64.

world in reverse. Actually, for the reader it is very difficult to understand Butler's point of view, because his predilection for paradoxical reasoning eventually mixes the parodied element (Victorian England) with the parody (Erewhon). [10] The relationship that Butler establishes with the protagonist, the traveller visiting Erewhon, is also ambiguous: sometimes he retreats from Higgs and criticises him, sometimes Higgs becomes the mouthpiece to express his ideas. In the same way, the traveller, who should be criticising Mr Nosnibor, the representative of the Erewhonians, acts according to the law of personal profit and both comply with social conventions.

A sort of neglected recluse during his life, Butler started being reassessed after his death. In the twentieth century many writers and thinkers recognised the importance of his works in their intellectual development. For G. B. Shaw [11] and Virginia Woolf [12] the incongruity of his style and ideas, far from being a limit, becomes a powerful weapon to criticize his own historical period. In order to deconstruct Victorian values and institutions, Butler uses the same tools that his Victorian education had endowed him with: irony and double-edged language. In this sense there is a close link between his style and his unstable, convoluted personality. In his *Notebooks* he liked to call himself 'unorthodox and militant'. So, paradoxically, in order to strike institutions and dogmatic certainties he uses the same logic and style he criticises in the Colleges of Unreason. Indeed, the places in Erewhon where knowledge is transmitted and hypothetical language is taught are a corrosive satire of the Victorian university system. [13]

The discursive strategies which find an explanation in Butler's own Victorian education and his being at the same time inside and outside society make his writings ambiguous and not easy to decipher. He is trapped within his own oppositional strategies and many of his intuitions, especially in his book on organic memory, do not manage to find an adequate development, because of the very passion he shows in contradicting them. Butler's language, double, paradoxical, chiastic, [14] and oxymora-filled as it is, comes from the necessity of contradiction, more than from a passion for symmetry. He is thus a sort of double-faced Janus: his anti-dogmatism is modern and projects him into the

---

10 Beatrice Battaglia, 'Erewhon', in *Histoire transnationale de l'utopie littéraire et de l'utopisme*, coordonnée par Vita Fortunati et Raymond Trousson, avec la collaboration de Paola Spinozzi (Paris: Champion, 2008), pp. 645–60.

11 George Bernard, Shaw, 'The New Life Reviewed', in *Pen, Portraits, and Reviews* (London: Constable and Company, 1932), pp. 52–71.

12 Virginia Woolf, 'A Man with a View' (1916), in *Contemporary Writers* (London: Hogarth Press, 1965), pp. 28–35.

13 Maria Luisa Bignami, 'Erewhon', in *Dictionary of Literary Utopias*, ed. by Vita Fortunati and Raymond Trousson (Paris: Champion, 2000), p. 205.

14 Ralf Norman, *Samuel Butler and the Meaning of Chiasmus* (London: Macmillan, 1986).

twentieth century, while his views about heredity and his attack against the Darwinian idea of natural selection appear still tied to the past.

III. The Darwin-Butler Controversy

The controversy that Samuel Butler engaged against Charles Darwin spotlights how the *Origin of Species* became, from its first appearance in 1856, the centre of discussion and strife in Great Britain and Europe. Indeed, the theory of evolution unhinged some of the principles on which the Victorian system was founded: it went against the idea that at the origin of the world there had been creation and a creating God, and especially against the idea that the species had been created in a supernatural way, by means of some inexplicable divine act at the very beginning, or through multiple geological epochs. For the Victorians the concept of chance variations, lacking a preordained plan, was unacceptable. This debate heralded controversies which are still raging between supporters of evolution and creationists, who believe in an intelligent design. Going back to the origin of this debate enables us to understand the reasons why Darwin's message was and still is so shocking in Western society. Although Butler's theories are in many ways anachronistic (they were so defined by Darwin's successors, T. H. Huxley and George Romanes), they had the merit of laying out a series of questions that Darwin himself had left unanswered because, as Peter Raby rightly points out:

> so little was known about genetics, there was no way of demonstrating how species might develop by means of variation within the grand design laid down by Darwin; and it was not until the application of Mendelian theory and, recently, the discovery of DNA and RNA that the general tenor of Darwin's theory could be firmly substantiated. [15]

The battle that Butler engaged against Darwin lasted for many years and in his book *Unconscious Memory* he reported each phase punctiliously. *Life and Habit* already exhibits some of the characteristics present in his later works: the argumentative, sometimes aggressive vein accompanies his love for abstruse speculation. It has rightly been noted that Darwin and his theory of evolution became somewhat of an obsession for Butler, causing him stress and psychosomatic disorders, such as, for instance, difficulty in breathing. 'Writing *Life and*

15 Raby, p. 168. On the other hand, as is shown also by Richard Dawkins's book, *The Blind Watchmaker* (1986), especially the chapter 'Doomed Rivals', the debate on Lamarck's theories continued also in the second half of the twentieth century.

Habit literally took my breath away. I kept wanting to take a long breath, and quite unable to do so'. [16]

What did Darwin represent for Butler? He read Darwin's *Origin of Species* in 1860–61 after having moved to New Zealand against his father's wishes, which were to destine him to the Church. Young Butler was searching for a new philosophy and new beliefs, wanting to escape those of the Catholic Church. He read the book in physical and mental solitude, in his estate in New Zealand, and immediately realised its importance. He claimed more than once that it was a fundamental stage in the development of his own ideas. If his attitude towards Darwin was at the beginning one of reverential fear and respect, later on he started to discuss the theory of evolution and in particular the concept of natural selection. Darwin became the symbol of authority and he ended up identifying him with his father's generation, whose power he liked defying. [17] If for a long period in Butler's life his father was his target, and with him religion, later on it was Darwin, and in particular the adoration the scientific world felt for this great scientist who, by becoming a sort of super-icon, turned into Butler's foe. The *Origin of Species*, ironically called 'The Book', was a sort of Bible, which, for this very reason, could not be considered infallible, but had to be approached with an open spirit, in order to answer a series of questions which Darwin had left unanswered and unexplored. The controversy can be understood only if one keeps in mind the main features of Butler's complex and intricate personality, that of a loner who lived on the margins of the cultural and scientific world of his time. From the details of this controversy there emerges the juxtaposition between the world of powerful scientists such as Darwin and his friends Huxley and Romanes, and Butler's own peculiar world, that of an eccentric writer who lived suffering because of his homosexuality and did not manage to obtain recognition for his works during his life. A writer who was reputed to be a bizarre character, an amateur who, driven by a sort of universal dilettantism, wanted to dedicate his attention to painting, photography, music, science, biology and literature. In short, someone interested in matters he had only an imperfect knowledge of. Darwin himself never commented on the quarrel with Butler and abstained from replying to his accusations of plagiarism. Moreover, the two families had been close, since Butler's grandfather had taught Charles Darwin at Shrewsbury School, [18] as Butler himself emphasised. The controversy thus conceals, especially on Butler's part, an involvement which was not only intellectual, but also emotional and psychological. What Butler found particularly

16 *Letters between Samuel Butler and Miss E. M. A. Savage 1871–1885*, ed. by Geoffrey Kenyes and Brian Hill (London: J. Cape, 1935), p. 190.
17 Phyllis Greenacre, *The Quest for the Father* (New York: International University Press, 1963).
18 Nora Barlow, 'Appendix', in *The Autobiography of Charles Darwin, 1809–1882: With Original Omissions Restored*, ed. by Nora Barlow (London: Collins, 1958), p. 175.

offensive was the general indifference with which the scientific world as a whole, and Darwin in particular, greeted his books. This explains the vehemence and the doggedness Butler showed in perfecting his theories, in the hope of promoting his ideas as well as receiving an adequate answer from the scientific world.

In the second chapter of *Unconscious Memory* Butler explains the reasons why he took up the study of evolution: 'It was impossible, however, for Mr. Darwin's readers to leave the matter as Mr. Darwin had left it. We wanted to know whence came that germ or those germs of life which, if Mr. Darwin was right, were once the world's only inhabitants'. [19] Butler's contention is based on the following points: [20] Darwin never quotes his predecessors, Buffon, Lamarck and especially his own grandfather, Erasmus Darwin. According to Butler, then, 'candid' Darwin was guilty of not acknowledging the work of the scientists who came before him. In his second book on evolution, *Evolution, Old and New; or, the Theories of Buffon, Dr Erasmus Darwin, and Lamarck, as compared with that of Mr Charles Darwin*, Butler demonstrated that Darwin's theories were based on the works of Buffon, Erasmus Darwin and Lamarck: by quoting the various editions of the *Origin of Species*, he claimed that Darwin had been extremely unwilling to recognise his debt to scientists who had significantly contributed to the development of the theory of evolution. In particular, Butler highlights how Darwin, even in the 'brief but imperfect historical survey' subsequently added to the book, did not recognize the importance his grandfather had had in the formulation of the theory. On the other hand, Butler's motivations reveal that he was incapable of grasping the difference between Darwin's theories and all preceding ones, because he did not accept the basis of evolutionary theory, that is natural selection. The only answer poor Butler obtained from Darwin is mentioned in *Unconscious Memory*. In the preface to the English translation of the life of Erasmus Darwin written by the German scientist Dr. Krause there is a statement which Butler attributed to Darwin: 'Erasmus Darwin's system was in itself a most significant first step in the path of knowledge which his grandson has opened up for us, but to wish to revive it at the present day, as has actually been seriously attempted, shows a weakness of thought and a mental anachronism which no one can envy'. [21] This sentence ignited Butler's violent reaction: in the first chapters of *Unconscious Memory* he rejected the principle of chance in natural selection and introduced the *sense of need* which had been defined in Lamarckian terms as 'le sentiment intérieur'. This term identified an element

19 Samuel Butler, *Unconscious Memory* (Teddington: The Echo Library, 1880; repr. 2006), p. 7.
20 For the controversy between Butler and Darwin see Basil Willey, *Darwin and Butler: Two Versions of Evolution* (London: Chatto and Windus, 1960); G. D. H. Cole, *Samuel Butler* (London: Longmans, Green, and Co., 1971).
21 *Unconscious Memory*, p. 22.

present in all beings, allowing them to respond to environmental stimuli and thus adapt. Butler describes the sense of need 'as faith or as desire to know, or do, or live at all'. [22] This vital force is directly connected to the idea of *unconscious memory* and a close link is established between hereditary and memory phenomena. Moreover, *unconscious memory* should be the common denominator of all phenomena present in the development of organic beings; the concept of *cunning* is juxtaposed to the concept of *luck*, which Butler illustrated in his last book on evolution *Luck or Cunning*, published in 1886. [23] The sense of need is an internal drive which makes every being, from the simplest to the most complex, tend to develop. Butler thus introduces the notion of finalism in the doctrine of evolution: life evolves not as the effect of mere chance, as Darwin claimed, but because in its infinite manifestations it possesses and reveals an inborn determination to evolve. Butler introduces a vaguely pantheistic conception which enables him to establish that above the evolution of biological species there is a teleological order of nature considered as a harmonic and animated whole. After having destroyed his Christian faith, Butler converted to vitalism. [24] So he substituted the idea of instant, once-for-all creation with a belief in the essential unity of life, that is in life with a sense of will, purpose and progress. The mechanical alternative, blind, random, soulless, as he saw it, was too awful to contemplate. Thus, Darwin's idea that life forms develop through 'the natural selection of numerous successive, slight, favourable variations' [25] was replaced by an emphasis on the sense of need aided by memory:

> Our own progress – or variation – is due not to small, fortuitous inventions or modifications which have enabled their fortunate possessors to survive in times of difficulty, not, in fact, to strokes of luck (though these of course, have had some effect – but no more, probably, than strokes of ill luck have counteracted) but to strokes of cunning – to a sense of need, and to study of the past and present which have given shrewd people a key with which to unlock the chambers of the future. [26]

According to Butler, this habit-forming process can be transmitted continuously from one generation to the next. He tried to show that this happened, and was of primary importance, by offering the theory of unconscious memory, which was

22 *Life and Habit*, p. 129.
23 Samuel Butler, *Luck, or Cunning as the Main Means of Organic Modification?* (Charleston, SC: BiblioBazaar, 1922; repr. 2008).
24 See Robert F. Rattray, 'The Philosophy of Samuel Butler', *Mind*, New Series, 23, 91 (July, 1914), 371–85 (p. 373).
25 Charles Darwin, *On the Origin of Species by Means of Natural Selection* (Chicago: Benton, 1978), p. 239.
26 *Life and Habit*, pp. 202–03.

part of each human's biological inheritance and could be transmitted, enriched by newly formed habits.

Then came one who told me that the stone was not mine, but that it had been dropped by Lamarck, to whom it belonged rightfully, but who had lost it; whereon I said I cared not who was the owner, if only I might use it and enjoy it. [27]

IV. The Theory of Organic Memory in *Unconscious Memory*

In her study on organic memory Laura Otis explains the reasons why it became a widespread and popular theory in Europe in the second half of the nineteenth century. [28] It has recently re-emerged during the Balkan wars, when dangerous nationalistic revanchist movements and issues of cultural identities rekindled racial prejudice, justifying horrific ethnic cleansing.

The relevance of Butler's theory to current issues consists in the close link he establishes between the hereditary processes of an individual and the development of a nation. In nineteenth-century Europe the theory of organic memory spreads because it not only supports rising nationalisms, but also the imperial ideology claiming the superiority of the white man on all coloured peoples. The popularity of this theory can be explained, then, on the basis of political and ideological reasons. As a matter of fact, the theory of organic memory offered a scientific basis to all those nations which emphasised the value of the features attributed to every ethnic group by linking the heredity of individuals to that of their nation. Furthermore, the second half of the nineteenth century was dominated by the fascination for origins. Such fascination permeated not only the European nations which, in order to define their boundaries, located them spatially, but also most knowledge disciplines, from philology to biology, from history to folklore, from geology to zoology. [29]

The theory of organic memory looked like a model which could be applied not only to the single individual but to society as a whole. Moreover, thanks to the analogical method, it became a useful tool for more than one discipline. As Otis aptly observes: 'It was because of analogy that the organic memory theory promised to unify all knowledge into a continuous system, and this aspect of the

27 Ibid., p. 250.

28 Laura Otis, *Organic Memory. History and the Body in the Late Nineteenth and Early Twentieth Centuries* (Lincoln and London: University of Nebraska Press, 1994).

29 On the origins of nations see *Mythen der Nationen: ein europäisches Panorama*, ed. by Monika Flacke (München – Berlin: Deutsches Historisches Museum, 1998; repr. 2001).

idea particularly appealed to lay thinkers writing on the outskirts of the scientific community'.[30]

The idea that every form of knowledge was part of a *continuum* is inherited from a Romantic conception of the world, according to which there is a close connection between the microcosm and the macrocosm. The analogical method could work, because the starting point was the definition of memory as the capacity to accumulate and reproduce past impressions: this capacity was possessed by the nervous system, by families and by languages. On the other hand, it is interesting to dwell on the meaning of the adjective *organic*, because this adjective is used to stress how memory is placed in the body, and in particular in the nervous system: it no longer belongs to the metaphysical, but to the physical and organic sphere, and thus to the domain of knowledge.[31] Organic memory presupposes a holistic vision in which the microcosm and the macrocosm are linked together. This concept is the foundation of Natural Philosophy in German Romantic culture, where the individual is the epitome of the universe.[32]

Thus, the body of an individual allows for understanding the development of its species and culture. The body must thus be considered as a palimpsest on which the traces of its interaction with the habitat, together with those of its family, and those of its ancestors, would accumulate.[33] The theory of organic memory raises a series of crucial ethical questions: how free are men in their choices and how much are they conditioned by family heritage?

It is one against legion when a creature tries to differ from his own past selves. He must yield or die if he wants to differ widely, so as to lack natural instincts, such as hunger, or thirsts, or not to gratify them. [...] His past selves are living in unruly hordes within him at this moment and overmastering him. 'Do this, this, this, which we too have done, and found our profit in it', cry the souls of his forefathers within him. [...] our former selves fight within us and wrangle for our possession. [...] Have we not here what is commonly called an *internal tumult*, when dead pleasures and pains tug within us hither and thither? Then may the battle be decided by what people are pleased to call our own experience. Our own indeed! What is our own save by mere courtesy of speech? A matter of fashion.[34]

30 *Organic Memory*, p. 19.

31 See note 67 below.

32 On the myths of German and English Romanticism see Anthony John Harding, *The Reception of Myth in English Romanticism* (Columbia and London: University of Missouri Press, 1995), and George S. Williamson, *The Longing for Myth in Germany. Religion and Aesthetic Culture from Romanticism to Nietzsche* (Chicago and London: The University of Chicago Press, 2004).

33 On this topic see Greta Perletti, *Le ferite della memoria. Il ritorno dei ricordi nella cultura vittoriana* (Bergamo: Bergamo University Press, 2008).

34 *Life and Habit*, pp. 43 – 44.

What weight does tradition have? How much space is left to artists to express their individuality? Precisely because Butler deeply believed in the strength of heredity, the real genius was he who possessed not only the capacity of assimilating, but also that of transforming what he inherited from tradition. The Romantic idea of a totally original creation thus vanishes. [35]

V. The Origin of Butler's Biological Discourse

The relevant role that tradition plays in the formation of individuality reflects, on a more general level, the influence that cultural heritage has on genetic inheritance. As frequently happens with Butler, the layers of discourse are reciprocally nested in a sort of *structure en abyme*. For this reason, to discuss the origin of his evolutionary discourse means to follow how he traced the original concept of evolution back to its forefathers.

As already mentioned, *Erewhon*, apart from being a dystopian satire of Victorian mores, laid down the foundations of Butler's biological discourse. A cursory look at how he came to write the novel will offer important hints on its inspirational source. In quite a peculiar and somehow innovative fashion, his prolific writing relies on a sort of modular technique, consisting in making up a book by assembling together and fleshing out previous, independent works. The peculiarity of his compositional method is certainly evident in his mature essays such as, for instance, *Unconscious Memory*, [36] but also in his earliest novel *Erewhon*, which seems to take shape from a few articles on Darwinian evolution. As admitted by Butler himself in *Erewhon*'s preface:

> The first part of *Erewhon* written was an article headed *Darwin among the Machines* [...]. It was written [...] in New Zealand, and appeared at Christchurch in the Press Newspaper, June 13, 1863. [...] A second article on the same subject as the one just referred to appeared in the Press shortly after the first, but I have no copy. It treated *Machines* from a different point of view, and was the basis of pp. 270–274 of the present edition of *Erewhon*. This view ultimately led me to the theory I put forward in *Life and Habit*, published in November 1877. [...] The next part of *Erewhon* that I wrote was the 'World of the Unborn', [which] appeared in some London paper [...], not very long after July 1, 1865, but I have no copy. I also wrote about this time the substance of what ultimately became the *Musical Banks*, and the trial of a man for

35 Robert Macfarlane, *Original Copy. Plagiarism and Originality in Nineteenth-Century Literature* (Oxford: Oxford University Press, 2007).

36 In the preface Butler states that the great part of the thirteen chapters of *Unconscious Memory* collects 'paralipomena' of his previous anti-Darwinian texts, like comments, jottings, epistles, or translation of passages on the unconscious matter from other memory-oriented scholars.

being in a consumption. These four detached papers were, I believe, all that was written of *Erewhon* before 1870. [37]

Such a list, which clearly reveals his already established Darwinian interests, fails to mention two other seminal works. *A First Year in Canterbury Settlement* (1863), [38] which is the literary account on his four-year colonial experience in New Zealand (1859 – 1864), can be regarded as the germ of his narrative technique, characterised by a 'contrapuntal style' and 'antipodean satire'. [39] *Lucubratio Ebria* (1865), sensibly the unmentioned second article which, Butler declared, had 'no copy', propounds the hypothesis that machines are 'extra-corporaneous limbs' [40] and can be considered his first step towards the teleological interpretation of evolution. [41] *Lucubratio Ebria* and *Darwin among the Machines*, written within a year's span (1863 – 4) as a sort of response to the *Origin of Species*, push the theory of natural selection to its extreme consequences so as to embrace the evolution of man-built contrivances: 'man has played that part among machines which natural selection has performed in the animal and vegetable kingdoms'. [42] Nevertheless, following his analogical reasoning, [43] based on the equation between the mind-guided evolution of machines operated by men and the chance-driven modification of organisms performed by natural selection, Butler blurred the fundamental distinction between the purposeful

37 Samuel Butler, 'Preface to Second Edition', in *Erewhon* (London: Penguin, 1872; repr. 1985), p. 29.

38 James G. Paradis, 'Introduction', in *Samuel Butler. Victorian against the Grain*, p. 8.

39 Roger Robinson, 'From Canterbury Settlement to Erewhon: Butler and Antipodean Counterpoint', in *Samuel Butler. Victorian against the Grain*, pp. 21 – 44.

40 Samuel Butler, *Lucubratio Ebria* (The Press of Christchurch, New Zealand, 1865), reprinted in Samuel Butler, *The Note-Books of Samuel Butler*, ed. by Henry Festing Jones (London: A. C. Fifield, 1918), pp. 47 – 53.

41 Hans-Peter Breuer, 'Samuel Butler's 'The Book of the Machines' and the Argument from Design', *Modern Philology*, 72 (1975), 365 – 83.

42 Samuel Butler, *Darwin among the Machines* (The Press of Christchurch, New Zealand, 13 June, 1863); reprinted in Samuel Butler, *A First Year in Canterbury Settlement, with Other Early Essays*, ed. by R. A. Streatfeild (London: Fifield, 1914), p. 181.

43 In his use of mechanism-organism analogy, Butler, like Darwin, was well aware of William Paley's *Natural Theology* (1802), where the teleological argument for the existence of God was first epitomized by the watchmaker analogy: looking at the complex mechanism of a watch one can find the evidence for the presence of a design and then of an intelligent designer; similarly, one can find evidences for the existence of a divine demiurge just looking at the complexity of organisms. As Breuer pointed out, another important source of inspiration for analogical reasoning was Bishop Joseph Butler (not a relative), who in the *Analogy of Religion, Natural and Revealed to the Constitution and Course of Nature* (1736) drew a series of general analogies between the principles of divine government and man's knowledge of nature (see Breuer, 'Samuel Butler's 'The Book of the Machines'', pp. 370 – 78). More recently, in *The Blind Watchmaker* Dawkins resets the old argument of intelligent design in the light of modern technology, by substituting the watch with complex computers.

and non-purposeful view of evolution. He then softened the idea that natural selection should be regarded as the sole explication of adaptive process, making the first step towards the Lamarckian perspective he was later to acquire.

Certainly important for introducing the teleological issue, these essays are also crucial for sketching out themes like the continuity between organic and inorganic matter and the power of volition on organismal adaptation; these themes will constitute the core of *Erewhon*'s biological speculations in the three *Book of the Machines* chapters [44], as well as the pillars of Butler's later theory of organic memory. By applying the materialistic notions he used to explain the emergence of organic life from inorganic compounds, he developed a materialistic view of the human mind and suggested that the substrate of memories is made of organic matter properly organized into brain cells. [45] In such a linear representation of the *scala naturae*, which echoes his Romantic idea of life as a *continuum* from inanimate matter to human consciousness, Butler assigned machines the significant role to show 'the ultimate development of mechanical consciousness'. [46] For him, machines were used by human beings not only as supplementary limbs for hard work, but they were also treated as living organisms and subject to man's selection, like domestic animals. The effect of breeding on time scale is to gain a mutation rate much faster than the one operating on animals by means of natural selection, the former depending on an *ad hoc* selection of certain aspects of machines, the latter following a random, trial-and-error process. Accordingly, Butler envisions a near future in which machines are supposed to be 'destined to supplant the race of man' by means of a 'descent of conscious (and more than conscious) machines from those which now exist'. [47] Not only, then, did machines represent the interface between inorganic and organic world, between mechanical and conscious matter, but they were also instrumental in demonstrating that the purposive action of men could direct the process of evolution.

What emerges from these early essays and will be better defined in *Erewhon* is a concept of evolution characterized both by a self-willed acquisition of traits and an indissoluble continuity among life forms. Although both hypotheses were also present in Darwin's *Origin of Species*, as shown, respectively, by the theory of use and disuse (subsequently elaborated as the theory of pangenesis [48])

44 The chapters entitled 'The Book of the Machines' are the following: XXIII, XXIV, XXV.

45 Samuel Butler, 'Darwin among the Machines', in *A First Year in Canterbury Settlement*, p. 181. See also Rattray, p. 381.

46 *Erewhon*, p. 199.

47 Ibid., pp. 97, 223, 202.

48 *The Variation of Animals and Plants under Domestication*, published in 1868, nine years after the *Origin of Species*, adopts a 'provisional theory of pangenesis' to account for the complex phenomenon of heredity. In its essence, the theory assumes the existence of hereditary

and that of common ancestry, Butler pushed them to their extreme consequences in order to corroborate his still-developing conceptualization of evolutionary identity. The formation of selfhood, with the burdensome plurality of selves, was both the starting point and final destination of Butler's biological discourse. It is not to be forgotten that his idea of personal identity extends across generations, being a result of an endless series of repeated uses – which in the course of time become habits and eventually unconscious memories – transmitted generation by generation from the original ancestor to the final offspring. Containing the collected memories of its ancestors, each individual organism experiences the weight of the past selves which form its evolutionary history, as the case of rose-seed presented in *Erewhon* clearly exemplifies:

> Did the rose-seed ever form part of the identity of the rose-bush on which it grew? [...] Was this rose-bush ever linked by all those links that we commonly consider as constituting personal identity, with the seed from which it in its turn grew? [...] The rose-seed did what it now does in the persons of its ancestors – to whom it has been so linked as to be able to remember what those ancestors did when they were placed as the rose-seed now is. Each stage of development brings back the recollection of the course taken in the preceding stage, and the development has been so often repeated, that all doubt – and with all doubt, all consciousness of action – is suspended. [...] How do you show that it actually did remember? By the action which each generation takes [...] which is explicable on [...] the supposition that there is an abiding memory between successive generations.[49]

Even at this embryonic stage of reflections on evolution, Butler reveals what will remain the general aim towards which all his apparently divergent biological arguments would converge: the quest for identity and the memory of its origins. Like many other Victorian writers, Butler was attracted by the social and familial formation of identity. Influenced by the tensions between the national identity of

particles (*gemmules*), which are constantly produced by the cells of all body (pan-genesis, that is the genesis of the whole) and are able to circulate throughout the whole organism and be modified by environmentally induced changes (acquired traits) during the different phases of development. When such particles leave the somatic cells (the body) to reach the germinal cells (the sperm and the ovum), the latter acquire the modified trait and transmit it to the genetic constitution of the offspring. Continuity (vs. segregation) between germinal and somatic cells, namely the exact mechanism by which acquired characters could be inherited, was the trait that Darwin's theory of pangenesis shared with Lamarck's theory of acquired characteristics. Few years after Darwin's proposal, the very same cellular mechanism was applied to nerve cells to explain the faculty of memory. See Alfred Sanders, 'Mr. Darwin's Hypothesis of Pangenesis as Applied to the Faculty of Memory', *Journal of Anthropology*, 1 (1870), 144–49; Id., 'On Mr. Darwin's Hypothesis of Pangenesis as Applied to the Faculty of Memory [Abstract and Comments]', *Journal of the Anthropological Society of London*, 8 (1870), cxlvi–clii.

49 *Erewhon*, pp. 239–40.

an expanding colonial empire and the social identity of industrial middle-class, Victorian selfhood was fascinated by the control of the future progress of society as well as by the (re)definition of its origins. [50] Butler's cultural context was then strongly characterized by fascination with the origins, a feeling which spans from the need to belong to a specific ethnic group to the rise of a more general interest towards ancestors. [51] But besides being a humanist, he was also a lay biologist, something which permitted him to add a fresh dimension to his approach to the origin of identity: the long-term genealogy. For the Butler of *Erewhon*, then, individuality was important in driving the evolution of domestic organisms (machines included) but also indebted to evolution for shaping the brain. His humanistic and scientific knowledge enabled him to describe the cultural and biological determinism underlying the process of identity formation.

VI. The Biological Background of Butler's Organic Memory Theory

By the mid-1870 Butler was already developing a reasonable explanation for what he had vaguely called 'an abiding memory between successive generations'. [52] In that period he was working on two texts, *The Way of All Flesh* and *Life and Habit*, respectively an autobiographical *Bildungsroman* and an evolutionary essay exploring the influence of evolutionary history on the formation of

50 In order to understand the cultural climate in which Butler developed his ideas about the future order of British civilization, it should be recalled that only few months before the publication of *Erewhon*, Edward Bulwer-Lytton's novel *The Coming Race* (1871) appeared. With Butler's dystopian novel, *The Coming Race* shares so many themes that the two books were believed to belong to the same author and to have reciprocally contributed to their success (see Butler's 'Preface to the Second Edition', in *Erewhon*, p. 29). Notably, both books established a sort of circularity between the expectation for a future society transformed (ameliorated?) by technology and its original antediluvian 'pure' race, believed from *The Coming Race*'s hero to be 'descended from the same ancestors as the great Aryan family'. See E. G. Bulwer Lytton, *The Coming Race* (S.l.: Edinburgh, 1871), Chapter XXVI.

51 In that period numerous publications on this topic were published: George Drysdale's *The Elements of Social Science. Physical, Sexual, and Natural Religion* (1854), Isabella Duncan's *Pre-Adamite Man* (1860), Charles Lyell's *The Antiquity of Man* (1860), John Lubbock's *Prehistoric Times* (1865), Charles Darwin's *Origin of Species* (1859) and *The Descent of the Man* (1871). Consequently, Victorian enquiries into the origins have been the object of numerous critical studies. See for instance: Macfarlane, p. 71; Jenny Bourne Taylor and Sally Shuttleworth, *Embodied Selves: An Anthology of Psychological Texts, 1830–1890* (Oxford: Clarendon Press, 1998), p. xiv; Sally Shuttleworth, 'The Evolutionary Psychology and the Way of All Flesh', in *Samuel Butler. Victorian against the Grain*, p. 152.

52 *Erewhon*, p. 240.

personality. [53] Up to this point, his idea of memory was closely aligned with the Darwinian hypothesis of common ancestry: 'the course of any germ's development', Butler argued in *Erewhon*, is 'dictated by the habits of the germ from which it was descended, and of whose identity it had once formed part'. [54] However, his increasing conviction that the acquisition of new traits (structures or habits) were mainly due to a latent impulse of the will, urged Butler to reassess ancestral memories as a by-product of purposeful adaptation, introducing, *de facto*, a teleological perspective which collided with Darwin's concept of chance-driven Nature. When *Life and Habit* was nearly finished, Butler decided to turn his back on the great naturalist and pursue the project of reinstating the factor of design in evolution. He then decided to add two chapters, the third before last explicitly entitled 'Lamarck and Mr. Darwin', where he recovered the analogy between artificial and natural design and established animal will as the major cause of biological and social progress.

By the end of the 1870s Butler provided what was, for that time, a scientifically coherent explanation for the origin and evolution of selfhood: the theory of unconscious memory. He substantiated his idea by referring to three main theories: the Lamarckian theory of acquired characters, Ernst Haeckel's theory of recapitulation, and Ewald Hering's theory of organic memory. Butler believed that the organisms' sense of need drives their habits that, thanks to repetitions (use/disuse), are eventually stabilized into instincts (Lamarck's theory). By equating heredity and memory (Hering's theory), he also introduced a physiological mechanism to explain how the experiences stored in the brain during organismal life span could be transmitted to the progeny by modifying the reproductive cells (Lamarck's theory), and how every individual could encapsulate the experiences and memories of their ancestors and parents back to the primordial cell (Haeckel's theory). A close scrutiny of the three theories will reveal how Butler blurred the boundaries between biological and cultural memories. By interweaving individual development and evolutionary history, parental heritage and genetic heredity, conscious memory and instinctive memories, he eventually built up a grandiose theory, which linked personal identity to its origins.

As for Lamarckism, it has to be remembered that Butler's adhesion to the theory of acquired characters did not make him an anti-evolutionist. Opposed as

53 Sally Shuttleworth, 'The Evolutionary Psychology and the Way of All Flesh', pp. 143–69 (pp. 143–44); Thomas E. Hart, 'Desire under the Eucalyptus: Evolution in the Way of All Flesh', <http://web.mac.com/tehart/Jurassic_Rants/Welcome_files/Desire%20Under%20the%20Eucalyptus.pdf> [accessed 30 August 2009], pp. 1–12.

54 *Erewhon*, p. 236; see also Marcus Hartog, 'Introduction to Samuel Butler's Unconscious Memory', in Samuel Butler, *Unconscious Memory*, ed. by R. A. Streatfeild (London: A. C. Fifield, 1910), p. xii.

he was to the selective principle, he not only accepted the concept of evolution, but he devoted himself to reinstating its original formulation as advanced by Buffon, Erasmus Darwin, and Lamarck. Butler felt he shared many of the tenets Lamarck enunciated in *Philosophie Zoologique* (1809). In their gross descriptions, they are the following: 1) the formation of a new organ or habit is the result of a new need; 2) all the experiences acquired by the individual (parents) could be passed on to its offspring (child); 3) the more organs are used (or disused), the bigger (or smaller) they become.

Although the inheritance of acquired characteristics (second tenet) is generally regarded as the core of Lamarck's theory, at that time it was shared by several naturalists, the Darwinian theory of pangenesis being indeed just one case among others. [55] Its specific mechanism allowed Butler to explain, in cellular terms, how a deep stimulation is first memorized by the body cells and then passed on to germinal level through the impregnation of the ovum or embryo, a mechanism that, on a more general level, he used to explain how continuity of personality from parents to child was possible.

Concerning the first tenet, we have already outlined how the self-willed acquisition of traits was present in Butler's essays on machines well beyond his acquaintance of Lamarck's writings. What rather captured Butler's attention was Lamarck's persuasive teleological explanation of organismal mutations. Such a finalistic version of evolution was for Butler instrumental in establishing that unconscious volition, i. e. the primeval spark of individuality, is the very first drive which leads organisms to transform life experiences and habits into genetically fixed instincts. [56] Finally, from Lamarck, and even more from Lamarckians, Butler, apparently following a typical translational mistake from the French *besoin*, exploited the vagueness of the term and skilfully coined the expression 'unconscious need' to define contradictory terms like volitional and instinctual actions, present and ancestral memories. [57] Unconscious thus refers to a sort of conceptual no man's land where 'perfect ignorance and perfect knowledge are alike' and where 'extremes [...] meet and become indistinguishable from one another'. [58] Lamarck's third tenet of use and disuse [59]

55 Edward J. Pfeifer, 'The Genesis of American Neo-Lamarckism', *Isis*, 56 (1965), 156 – 67. See also the classic book of Peter J. Bowler, *The Eclipse of Darwinism: Anti-Darwinian Evolution Theories in the Decades around 1900* (Baltimore: Johns Hopkins University Press, 1983).

56 *Life and Habit*, p. 15.

57 See Rattray, p. 373; Beatrice Edgell, *Theories of Memory* (Oxford: Clarendon Press, 1924), p. 118; Thomas E. Hart, 'The Role of Will in Two Evolutionary Plays by Bernard Shaw' (dissertation, The Catholic University of America, 1993), p. 12, n. 8. A detailed discussion of the English misinterpretations of Lamarck's French terminology is also offered by Willey, pp. 46 – 47.

58 *Life and Habit*, pp. 8, 15.

59 By extending the equation between use and size from body structures to mental activities and

allows Butler to argue that through repetition all habitual psychological actions become unconscious: 'acts which we do without knowing that we do them'. [60]

The acquaintance of Ernst Haeckel's law of recapitulation did not represent a complete novelty for Butler's theory, but was rather a scientific corroboration of his already established idea that the patterns of individual life (ontogeny) mirror that of species (phylogeny). Theorized in *Generelle Morphologie der Organismen* (1866) as a law explaining the analogy of embryological development of organs in different species, Haeckel's law was soon extended by Butler beyond the womb and physical dimension to embrace psychological process and society evolution. [61] Far-reaching as it was, the idea of recapitulation soon gained the centre stage of his doctrine 'bringing [...] apparently conflicting phenomena under the operation of one law'. [62] Even more, such a theoretical framework enabled him to place the childhood of the race, ancestral memories, and the origin of personality on the same line of development: 'the continuation of life or personal identity in the individual and the race are fundamentally of the same kind, or, in other words, there is a veritable prolongation of identity'. [63]

By promoting the creation of prolific analogies between inherited ancestral memories and cultural heritage, Butler's recapitulation finally achieved the long desired union between the individual and its origins: personal achievements were dependent on familial traditions, individual moral depravity was better understandable as a disease inherited from weak ancestors. The development of the body cannot avoid the influence of the origins; as Rattray observes, 'the way in which we grow our bodies is, according to Butler, due to inherited memories of the very ancient kind'. [64] The latter point, which Haeckel developed a decade after the formulation of his law, stimulated further reflection. In a pamphlet entitled *Die Perigenese der Plastidule* (1875–6), Haeckel proposed a cellular mechanism to explain the transmission of the memories and stimuli acquired during the development from body organs to the reproductive cells. Even if this hypothesis appealed to Butler as the first attempt to embody life memories at organic level, it was another coeval theory which, like Haeckel's, came to his

behaviours and by replacing the enlargement of the size with the stabilization of habits, Butler was able to use the analogy as a sort of general principle, according to which the repeated use of a certain behaviour transforms habits into instincts and instincts into memories.

60 *Life and Habit*, p. 99.

61 For a detailed history of the theory of recapitulation and its adoption by humanistic disciplines like psychology and anthropology see Paul Mengal and Jean Céard, *Histoire du concept de récapitulation: ontogenèse et phylogenèse en biologie et sciences humaines* (Paris: Masson, 1993).

62 *Life and Habit*, p. 41.

63 *Unconscious Memory*, p. 34.

64 Rattray, p. 376.

attention thanks to Francis Darwin and struck his imagination to the point that he fully incorporated it in his general picture of unconscious memory. [65]

Shortly after its publication Butler read 'Memory as a Universal Function of Organised Matter', the Inaugural Address delivered by the prominent German physiologist Dr. Hering before the Imperial Royal Academy of Sciences at Vienna in 1870: in it he seemed to recognize assumptions different from his own. Like other Lamarckian-oriented approaches to cellular memory, Hering's theory of organic memory supported the doctrine of the transmission of body experiences to the hereditary apparatus. In his theory he rejected the existence of the physical barrier between genetic substance contained in the sperm and ovum (germ line) and the rest of the body (somatic line) [66] – or, as we could say nowadays, between genetic information and environmental stimuli –, hypothesized by the German physiologist August Weismann to support Darwin's idea that the accumulation of variations in species was aimless and 'fortuitous'. [67] Indeed, the novelty of organic theory, as Butler first recognized, consisted in stressing the communication between the body and the germ cell systems so as to envision a sort of identification between memory and heredity. What is more, he was deeply fascinated by the four ancillary hypotheses out of which such an identification was built, namely: the sound physical explanation of memory based on molecular vibration, the continuity between organic and inorganic matter, the centrality of the nervous system as a locus of unconscious memories, and the emphasis on origins.

Hering's theory suggested that when an organism comes across an environmental stimulus, a trace is left in his nerve fibers (somatic cells) by means of a specific vibration in the intra-cellular molecular disposition. The vibration or alteration persists within the cell and stimulates its growth, the persistence being allowed by the ability of the vibration to be transmitted both to the incoming nutritional molecules and then to the daughter cells. If such a vibration is often repeated, it eventually reaches the germ cells, altering their molecular disposition and then passing to the next generation. In such a physical view, memory is not a mere function of few specific parts of the nervous system, but an intrinsic characteristic of all organic (both in the sense of living and organized) matter. [68] Being a universal property of organic matter, 'memory must be re-

65 Hartog, p. xii.

66 *The Eclipse of Darwinism*, pp. 74, 82.

67 *Unconscious Memory*, p. 19.

68 Organic and organized are both included in the German term 'organisiert', which was originally used by Hering in his 1870 conference. See Bronislaw R. Gomulicki, *The Development and Present Status of the Trace Theory of Memory* (Oxford: Merton College, 1950), p. 9, n. 2; and Jacques Gasser, 'La notion de mémoire organique dans l'œuvre de T. Ribot', in *History and Philosophy of the Life Sciences*, 10, 2 (1988), 293–313. The influence of Butler's

garded as an unconscious, material process – a function of a brain function'.[69] Habits and motor reflexes in the nervous system, but also muscle reinforcement or cases of performance-reproduction in cells or in organs are all mnemonic phenomena which pass unperceived like many other unconscious biochemical processes. This explains how unconscious memories (involuntary motor reflex or latent instinct) could pass into conscious or organic memory (intentional movement or recollection) and vice versa.

Even if other physiologists and psychologists anticipated many elements of the organic theory,[70] Hering's proposal appeared to Butler the sole to have the great merit of intertwining a solid physical explanation of the mnemonic process with a monistic view of nature. Butler devoted paragraphs V and VI of his *Unconscious Memory* to translating and commenting Hering's theory: the major point is that the continuity which connects inorganic matter, hereditary mechanism and unconscious activity erases the boundaries between life and memory: 'Life is that property of matter whereby it can remember – matter which can remember is living'.[71] Such a circularity between life and memory also reinforces the temporal circularity between the actual individual and its origins. According to this idea, recollections and memories forming personal identity reflect, i. e. are the sum of, the history of the ancestors back to its origin, as Butler's translation of Hering's conclusion makes clear:

> We must ascribe both to the brain and body of the new-born infant a far-reaching power of remembering or reproducing things which have already come to their development thousands of times in the persons of their ancestors.[72]

What should now be clear is how and why Butler built his theory of organic memory. In his multi-layered conception of identity, he sees two interwoven levels of memories: the biological and the cultural one. At biological level, Haeckels's hypothesis of recapitulation helped him to establish a relationship between the development of individual life and the stages of its evolution. Out of the womb, the individual's personality becomes subject to the experiences of the family and its cultural heritage. These memories/experiences, although able to

---

organic theory during the resurgence of Lamarckian-oriented theories, and specifically of the trace or 'engram' theories of memory, in the first decades of the twentieth century has been investigated by Daniel L. Schacter in *Forgotten Ideas, Neglected Pioneers: Richard Semon and the Story of Memory* (Philadelphia – Hove: Psychology Press, 2001), in part. chapter VII.

69 *The Development and Present Status of the Trace Theory of Memory*, p. 9.

70 For an historical reconstruction see Bronislaw R. Gomulicki, Jacques Gasser, Beatrice Edgell, Ruth Parkin-Gounelas, and Sally Shuttleworth.

71 *Unconscious Memory*, p. 100.

72 Ibid., p. 50.

enrich each generation with new elements, are shaped by genetic traits which were formed, according to Lamarck's theory of use and disuse (i. e., repetitions), by conscious habits stabilized into unconscious instincts. Hering's organic theory finally offered Butler a cellular explanation for the reciprocal transformation of deliberate mnemonic process (memory) into unconscious organic process (heredity) (see Figure 8).

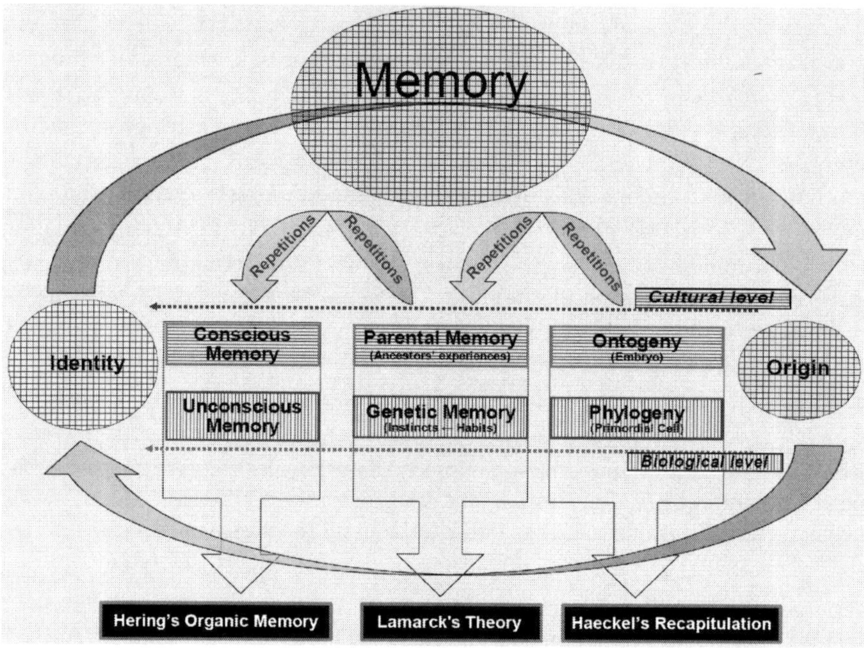

Figure 8: Samuel Butler's theory of unconscious memory.
The three anti-Darwinian theories were used to justify three different sets of memories constituting the identity of each individual. Each set is characterized by a biological and cultural level represented, respectively, by vertical and horizontal stripes. The two levels of biological and cultural memories influence each other within and across their own set thanks to the mechanism of repetitions, with the only difference that the more they depart from the origin, the more conscious they become. Being the result of the interplay of six different kinds of memory based on two levels, identity, memory and origin are represented by the overlappings between vertical and horizontal stripes.

VII. Conclusions

Butler's original attraction to evolutionary thought was inspired by his personal interest in man's dependence upon his origins. Shortly after reading the *Origin of Species*, he elaborated a personal interpretation of evolution whose tenets, centred on purposive adaptations, constituted the basis first of his texts on machines, and then of *Erewhon.* Such a teleological version of evolution became more and more radical owing to his increasing divergence from Darwin's natural selection principle. The emotional and cultural crisis that followed the abandonment of Darwin's theory engendered Butler's keen feeling of marginalization from mainstream biological thinking; this event made him aware of the importance of adhering to a specific profession, and of writing with a style suited to persuade the scientific community. It is in this context that he decided to underpin his vast portrait of organic memory with three 'scientific' anti-Darwinian theories.

Piqued as he was by the universal consensus surrounding Darwin, he preferred to support the previous generation of evolutionists and their notion of preordained design. Such a theoretical perspective eventually led him to overlook the powerful innovative character of natural selection: a process of animal adaptation to the environment where any organismal identity recounts its past evolutionary history, but whose path was written by blind chance and where the will is ineffective from a hereditary perspective.

Darwin's view of the future of mankind is not deterministic, whereas Butler's conception of ancestral memory leaves little room, if any, for free will or chance. Butler, indeed, was as iconoclastic in his style and critical against the structure of family and religion, as he was conservative in science. Furthermore, the effective modernity of ideas such as the multi-dimensional conception of identity and the anticipation of a machine-driven society should be reconsidered in the light of the fact that both prefigure a future which is a perpetuation of the past.

For Butler, then, the quest for identity is intertwined with the problem of cultural and evolutionary origins: any single individual personality is nothing but the result of ancestral, unconscious experiences which determine the individual's present, conscious choices. The fate of the individual is a path which one travels backwards, constantly looking back at the origins.

Bibliography Part I

Achilli, A., and others, 'Mitochondrial DNA Variation of Modern Tuscans Supports the Near Eastern Origin of Etruscans', *American Journal of Human Genetics*, 80 (2007), 759–68.

Achilli, A., and others, 'Saami and Berbers – An Unexpected Mitochondrial DNA Link', *American Journal of Human Genetics*, 76 (2005), 883–86.

Achilli, A., and others, 'The Molecular Dissection of mtDNA Haplogroup H Confirms that the Franco-Cantabrian Glacial Refuge was a Major Source for the European Gene Pool', *American Journal of Human Genetics*, 75 (2004), 910–18.

Aiken, N. E., *The Biological Origins of Art* (Westport: Praeger, 1998).

Ali, R. E., and S. I. S. Rattan, 'Curcumin's Biphasic Hormetic Response on Proteasome Activity and Heat Shock Protein Synthesis in Human Keratinocytes', *Ann. NY Acad. Sci*, 1067 (2006), 394–99.

Ammerman, A. J., and L. L. Cavalli-Sforza, *The Neolithic Transition and the Genetics of Populations in Europe* (Princeton: Princeton University Press, 1984).

Amorim, A., 'Archaeogenetics', *Journal of Iberian Archaeology*, 1 (1999), 15–25.

Anderson, S., and others, 'Sequence and Organization of the Human Mitochondrial Genome', *Nature*, 290 (1981), 457–65.

Andrews, R. M., and others, 'Reanalysis and Revision of the Cambridge Reference Sequence for Human Mitochondrial DNA', *Nature Genetics*, 23 (1999), 147.

Avital, T., 'Footprints Literacy: The Origins of Art and Prelude to Science', *Symmetry: Art and Science* (The Quarterly of the International Society for the Interdisciplinary Study of Symmetry), 9, 1 (1998), 3–46. Published also electronically by *Visual Mathematics – VISMATH*, The Electronic Supplement of *Symmetry – ISIS*, 1, 2 (1999), <http://members.tripod.com/vismath/> [accessed 15 August 2009].

Avital, T., 'The Origins of Art: An Archaeological or a Philosophical Problem?', *South African Journal of Art History*, 16 (2001), 34–57, <http://www.hit.ac.il/staff/Avital/ORIGINS.pdf> [accessed 15 August 2009].

Avital, T., *Art Versus Nonart: Art Out of Mind* (Cambridge – New York: Cambridge University Press, 2003).

Bandelt, H.-J., V. Macaulay, and M. Richards, *Human Mitochondrial DNA and the Evolution of Homo Sapiens* (Berlin – Heidelberg: Springer Verlag, 2006).

Barlow, N., ed., *The Autobiography of Charles Darwin, 1809–1882: With Original Omissions Restored* (London: Collins, 1958).

Bar-Yosef, O., 'The Upper Paleolithic Revolution', *Annual Review of Anthropology*, 31 (2002), 363 – 93.

Battaglia, B., '*Erewhon*', in *Histoire transnationale de l'utopie littéraire et de l'utopisme*, coordonnée par V. Fortunati et R. Trousson, avec la collaboration de P. Spinozzi (Paris: Champion, 2008), pp. 645 – 60.

Beer, G., *Open Fields: Science in Cultural Encounters* (New York: Oxford University Press, 1966).

Bessenyei, B., M. Márka, L. Urbán, M. Zeher, and I. Semsei, 'Single Nucleotide Polymorphisms: Aging and Diseases', *Biogerontology*, 5 (2004), 291 – 300.

Bignami, M. L., 'Erewhon', in *Dictionary of Literary Utopias*, ed. by V. Fortunati and R. Trousson (Paris: Champion, 2000), p. 205.

Bowler, P. J., *The Eclipse of Darwinism: Anti-Darwinian Evolution Theories in the Decades around 1900* (Baltimore: Johns Hopkins University Press, 1983).

Breuer, H.-P., 'Samuel Butler's 'The Book of the Machines' and the Argument from Design', *Modern Philology*, 72 (1975), 365 – 83.

Bulwer Lytton, E. G., *The Coming Race* (S.l.: Edinburgh, 1871).

Butler, S., *Darwin among the Machines* (The Press of Christchurch, New Zealand, 13 June, 1863); reprinted in S. Butler, *A First Year in Canterbury Settlement, with Other Early Essays*, ed. by R. A. Streatfeild (London: Fifield, 1914).

Butler, S., *Erewhon, or Over the Range* (London: Penguin, 1872; repr. 1985).

Butler, S., *Life and Habit* (New York: AMS Press, 1968).

Butler, S., *Luck, or Cunning as the Main Means of Organic Modification?* (Charleston, SC: BiblioBazaar, 1922; repr. 2008).

Butler, S., *Lucubratio Ebria* (The Press of Christchurch, New Zealand, 1865), reprinted in Samuel Butler, *The Note-Books of Samuel Butler*, ed. by H. F. Jones (London: A. C. Fifield, 1918).

Butler, S., *Unconscious Memory* (Teddington: The Echo Library, 1880; repr. 2006).

Butler, S., *Unconscious Memory*, ed. by R. A. Streatfeild with an Introduction by M. Hartog (London: A. C. Fifield, 1910).

Cann, R. L., M. Stoneking, and A. C. Wilson, 'Mitochondrial DNA and Human Evolution', *Nature*, 325 (1987), 31 – 36.

Carnes, B. A., S. J. Olshansky, and D. Grahn, 'Biological Evidence for Limits to the Duration of Life', *Biogerontology*, 4 (2003), 31 – 45.

Cavalli-Sforza, L. L., P. Menozzi, and A. Piazza, *The History and Geography of Human Genes* (Princeton: Princeton University Press, 1994).

Chalmers, F. G., 'The Study of Art in a Cultural Context', in *The Journal of Aesthetics and Art Criticism*, 32, 2 (Winter 1973), 249 – 56.

Cole, G. D. H., *Samuel Butler* (London: Longmans, Green, and Co., 1971).

Cooper, J. M., and D. S. Hutchinson, eds, *Plato. Complete Works* (Indianapolis – Cambridge: Hackett Publishing Company, 1997).

Coote, J., and A. Shelton, eds, *Anthropology, Art and Aesthetics* (Oxford: Clarendon Press, 1992).

Cordaux, R., and M. Stoneking, 'South Asia, the Andamanese, and the Genetic Evidence for an 'Early' Human Dispersal Out of Africa', *American Journal of Human Genetics*, 72 (2003), 1586 – 90.

Darwin, C., *On the Origin of Species by Means of Natural Selection, or the Preservation of Favoured Races in the Struggle for Life* (London: John Murray, 1859).
Darwin, C., *The Descent of the Man* (London: John Murray, 1871).
Darwin, C., *The Variation of Animals and Plants under Domestication* (London: John Murray, 1868).
Davis, W., 'Beginning the History of Art', *The Journal of Aesthetics and Art Criticism*, 51, 3 (Summer 1993), 327 – 50.
Dawkins, R., *The Blind Watchmaker* (London: Longman, 1986).
Dissanayake, E., *Art and Intimacy. How the Arts Began* (Seattle and London: University of Washington Press, 2000).
Dissanayake, E., *Homo Aestheticus: Where Art Comes From and Why* (Seattle and London: University of Washington Press, 1995).
Dissanayake, E., *What is Art For?* (Seattle and London: University of Washington Press, 1988).
Dolson, B., *Notes on The Origins of Nonrepresentational Art*, <http://www.billdolson.com/text/origins.htm> [accessed 15 August 2009].
Drysdale, G., *The Elements of Social Science. Physical, Sexual, and Natural Religion* (London: E. Truelove, 1854).
Duncan, I., *Pre-Adamite Man* (London – Edinburgh: Saunders, Otley & Co., 1860).
Dupanloup, I., and others, 'Estimating the Impact of Prehistoric Admixture on the Genome of Europeans', *Molecular Biology and Evolution*, 21 (2004), 1361 – 72.
Dutton, D., 'Aesthetics and Evolutionary Psychology', in *The Oxford Handbook of Aesthetics*, ed. by J. Levinson (New York: Oxford University Press, 2003).
Edgell, B., *Theories of Memory* (Oxford: Clarendon Press, 1924).
Farhat, M., 'Historical Constructions of Arab Art and Visual Culture', excerpted from the exhibition catalogue *Three Arab Painters in New York: Samia Halaby, Sumayyah Samaha and Athir Shayota*, copyrighted to M. Farhat and Al Jisser Group, May 2006, <http://www.arteeast.org/artenews/artenews-articles2006/historical-constr7-06.html> [accessed 15 August 2009].
Ferrari, C. K. B., 'Functional Foods, Herbs and Neutraceuticals: towards Biochemical Mechanisms of Healthy Aging', *Biogerontology*, 5 (2004), 275 – 89.
Finch, C. E., *Longevity, Senescence, and the Genome* (Chicago: The University of Chicago Press, 1990).
Flacke, M., ed., *Mythen der Nationen: ein europäisches Panorama* (München – Berlin: Deutsches Historisches Museum, 1998; repr. 2001).
Forster, P., and S. Matsumura, 'Evolution. Did Early Humans Go North or South?', *Science*, 308 (2005), 365 – 66.
Gasser, J., 'La notion de mémoire organique dans l'œuvre de T. Ribot', in *History and Philosophy of the Life Sciences*, 10, 2 (1988), 293 – 313.
Gomulicki, B. R., *The Development and Present Status of the Trace Theory of Memory* (Oxford: Merton College, 1950).
Gravina, B., P. Mellars, and C. B. Ramsey, 'Radiocarbon Dating of Interstratified Neanderthal and Early Modern Human Occupations at the Chatelperronian Type-site', *Nature*, 438 (2005), 51 – 56.
Green, R. E., and others, 'A Complete Neanderthal Mitochondrial Genome Sequence Determined by High-Throughput Sequencing', *Cell*, 134 (2008), 416 – 26.

Green, R. E., and others, 'Analysis of One Million Base Pairs of Neanderthal DNA', *Nature*, 444 (2006), 330 – 36.

Greenacre, P., *The Quest for the Father* (New York: International University Press, 1963).

Haak, W., and others, 'Ancient DNA from the First European Farmers in 7500-year-old Neolithic Sites', *Science*, 310 (2005), 1016 – 18.

Harding, A. J., *The Reception of Myth in English Romanticism* (Columbia and London: University of Missouri Press, 1995).

Harries, K., *Art Matters. A Critical Commentary on Heidegger's 'The Origin of the Work of Art'* (Dordrecht: Springer, 2009, Contributions to Phenomenology, vol. 57).

Hart, T. H., 'Desire under the Eucalyptus: Evolution in the Way of All Flesh', <http://web.mac.com/tehart/Jurassic_Rants/Welcome_files/Desire%20Under%20the%20Eucalyptus.pdf> [accessed 30 August 2009].

Hart, T. H., 'The Role of Will in Two Evolutionary Plays by Bernard Shaw' (dissertation, The Catholic University of America, 1993).

Hayes, D. P., 'Nutritional Hormesis', *Eur J Clin Nutr*, 61 (2007), 147 – 59.

Hayflick, L., 'Biological Aging Is No Longer an Unsolved Problem', *Ann. NY Acad. Sci*, 1100 (2007), 1 – 13.

Hayflick, L., 'Entropy Explains Aging, Genetic Determinism Explains Longevity, and Undefined Terminology Explains Misunderstanding Both', *PLoS Genet.*, 3 (2007), e220.

Heidegger, M., *Der Ursprung des Kunstwerkes* (Ditzingen: Reclam Universal-Bibliothek, 1986).

Heidegger, M., *Holzwege* (Frankfurt: Klostermann, 2003).

Herodotus, *The Histories*, c. 430 BC, *Book One*, 94, trans. with Notes by G. Rawlinson, With an Introduction by T. Griffith (London: Wordsworth, 1996), pp. 45 – 46.

Holt, L. E., 'Samuel Butler and His Victorian Critics', *ELH*, 8, 2 (June 1941), 146 – 59.

Hudjashov, G., and others, 'Revealing the Prehistoric Settlement of Australia by Y chromosome and mtDNA Analysis', *Proceedings of the National Academy of Sciences of the United States of America*, 104 (2007), 8726 – 30.

Jobling, M. A., M. E. Hurles, and C. Tyler-Smith, *Human Evolutionary Genetics: Origins, Peoples and Disease* (New York: Garland Science, 2004).

Johnson, T. E., K. Christensen, and J. W. Vaupel, 'The Quest for Genetic Determinants of Human Longevity: Challenges and Insights', *Nature Rev. Genet.*, 7 (2006), 436 – 48.

Keynes, G., and B. Hill, *Letters between Samuel Butler and Miss E. M. A. Savage 1871 – 1885* (London: J. Cape, 1935).

Kimmelman, M., 'ART; A Heart of Darkness in the City of Light', *Times*, 2 July 2006, <http://www.nytimes.com/2006/07/02/arts/design/02kimm.html> [accessed 15 August 2009].

Kirkwood, T. B. L., and T. Cremer, 'Cytogerontology since 1881: a Reappraisal of August Weismann and a Review of Modern Progress', *Hum. Genet.*, 60 (1982), 101 – 21.

Kirkwood, T. B. L., *Time of Our Lives* (London: Weidenfeld & Nicolson, 1999).

Kozlowski, J., and M. Otte, 'The Formation of the Aurignacian', *Journal of Anthropology Research*, 56 (2000), 513 – 24.

Krings, M., and others, 'DNA Sequence of the Mitochondrial Hypervariable Region II from the Neanderthal Type Specimen', *Proceedings of the National Academy of Sciences of the United States of America*, 96 (1999), 5581 – 85.

Kuningas, M., S. P. Mooijaart, D. van Heemst, B. J. Zwaan, P. E. Slagboom, and R. G. J.

Westendorp, 'Genes Encoding Longevity: from Model Organisms to Humans', *Aging Cell*, 7 (2008), 270 – 80.

Le Bourg, E., and S. I. S. Rattan, 'Can Dietary Restriction Increase Longevity in All Species, Particularly in Human Beings? Introduction to a Debate Among Experts', *Biogerontology*, 7 (2006), 123 – 25.

Le Bourg, E., and S. I. S. Rattan, eds, *Mild Stress and Healthy Aging: Applying Hormesis in Aging Research and Interventions* (Dordrecht: Springer, The Netherlands, 2008).

Levy, E. K., 'Contemporary Art and the Genetic Code: New Models and Methods of Representation', *Art Journal*, 55, 1 (Spring, 1996), 20 – 24.

Lewis-Williams, D., *The Mind in the Cave: Consciousness and the Origins of Art* (London: Thames & Hudson, 2002.

Lubbock, J., *Prehistoric Times* (S.l.: Williams & Norgate, 1865).

Lyell, C., *The Antiquity of Man* (London: John Murray, 1860).

Macaulay, V., and others, 'Single, Rapid Coastal Settlement of Asia Revealed by Analysis of Complete Mitochondrial Genomes', *Science*, 308 (2005), 1034 – 36.

Macfarlane, R., *Original Copy. Plagiarism and Originality in Nineteenth-Century Literature* (Oxford: Oxford University Press, 2007).

Martin, G. M., 'Genetic Modulation of Senescent Phenotypes in *Homo sapiens*', *Cell*, 120 (2005), 523 – 32.

Mellars, P., 'A New Radiocarbon Revolution and the Dispersal of Modern Humans in Eurasia', *Nature*, 439 (2006), 931 – 35.

Mellars, P., 'Going East: New Genetic and Archaeological Perspectives on the Modern Human Colonization of Eurasia', *Science*, 313 (2006), 796 – 800.

Mellars, P., 'Why Did Modern Human Populations Disperse from Africa ca. 60,000 years ago? A New Model', *Proceedings of the National Academy of Sciences of the United States of America*, 103 (2006), 9381 – 86.

Mellars, P., B. Gravina, and C. B. Ramsey, 'Confirmation of Neanderthal/Modern Human Interstratification at the Chatelperronian Type-site', *Proceedings of the National Academy of Sciences of the United States of America*, 104 (2007), 3657 – 62.

Mengal, P., and J. Céard, *Histoire du concept de récapitulation: ontogenèse et phylogenèse en biologie et sciences humaines* (Paris: Masson, 1993).

Menozzi, P., A. Piazza, and L. L. Cavalli-Sforza, 'Synthetic Maps of Human Gene Frequencies in Europeans', *Science*, 201 (1978), 786 – 92.

Mithen, S., *The Prehistory of the Mind: A Search for the Origins of Art, Religion, and Science* (London: Thames & Hudson, 1996).

Nei, M., and A. K. Roychoudhury, 'Evolutionary Relationships of Human Populations on a Global Scale', *Molecular Biology and Evolution*, 10 (1993), 927 – 43.

Noonan, J. P., and others, 'Sequencing and Analysis of Neanderthal Genomic DNA', *Science*, 314 (2006), 1113 – 18.

Norman, R., *Samuel Butler and the Meaning of Chiasmus* (London: Macmillan, 1986).

Olivieri, A., and others, 'The mtDNA Legacy of the Levantine Early Upper Palaeolithic in Africa', *Science*, 314 (2006), 1767 – 70.

Olshansky, S. J., and T. T. Perls, 'New Developments in the Illegal Provision of Growth Hormone for 'Anti-aging' and Bodybuilding', *JAMA*, 299 (2008), 2792 – 94.

Otis, L., ed., *Literature and Science in the Nineteenth Century. An Anthology* (Oxford: Oxford University Press, 2002).

Otis, L., *Organic Memory. History and the Body in the Late Nineteenth and Early Twentieth Centuries* (Lincoln and London: University of Nebraska Press, 1994).

Otte, M., 'The Northwestern European Plain around 18.000 BP', in *The World at 18.000 BP*, ed. by O. Soffer and C. Gamble (London: Unwin, 1990), pp. 54 – 68.

Paradis, J. G., ed., *Samuel Butler. Victorian against the Grain: A Critical Overview* (Toronto – London: University of Toronto Press, 2007).

Perletti, G., *Le ferite della memoria. Il ritorno dei ricordi nella cultura vittoriana* (Bergamo: Bergamo University Press, 2008).

Pfeifer, E. J., 'The Genesis of American Neo-Lamarckism', *Isis*, 56 (1965), 156 – 67.

Quintana-Murci, L., and others, 'Genetic Evidence of an Early Exit of *Homo Sapiens Sapiens* from Africa through Eastern Africa', *Nature Genetics*, 23 (1999), 437 – 41.

Raby, P., *Samuel Butler. A Biography* (London: The Hogarth Press, 1991).

Rattan, S. I. S., 'Ageing, Gerontogenes, and Hormesis', *Ind. J. Exp. Biol*, 38 (2000), 1 – 5.

Rattan, S. I. S., 'Anti-ageing Strategies: Prevention or Therapy?', *EMBO Reports*, 6 (2005), S25 – S29.

Rattan, S. I. S., 'Biogerontology: the Next Step', *Ann. N.Y. Acad. Sci.*, 908 (2000), 282 – 90.

Rattan, S. I. S., 'Increased Molecular Damage and Heterogeneity as the Basis of Aging', *Biol. Chem*, 389 (2008), 267 – 72.

Rattan, S. I. S., 'The Science of Healthy Aging: Genes, Milieu, and Chance', *Ann N Y Acad Sci*, 1114 (2007), 1 – 10.

Rattan, S. I. S., 'Theories of Biological Aging: Genes, Proteins and Free Radicals', *Free Rad. Res*, 40 (2006), 1230 – 38.

Rattan, S. I. S.,'Gerontogenes: Real or Virtual?', *FASEB J.*, 9 (1995), 284 – 86.

Rattan, S. I. S., 'Hormesis in Aging', *Ageing Res. Rev.*, 7 (2008), 63 – 78.

Rattray, R. F.,'The Philosophy of Samuel Butler', *Mind*, New Series, 23, 91 (July, 1914), 371 – 85.

Richards, M., and others, 'In Search of Geographical Patterns in European Mitochondrial DNA', *American Journal of Human Genetics*, 71 (2002), 1168 – 74.

Rose, M. R., *Evolutionary Biology of Aging* (New York: Oxford University Press, 1991).

Sanders, A., 'Mr. Darwin's Hypothesis of Pangenesis as Applied to the Faculty of Memory', *Journal of Anthropology*, 1 (1870), 144 – 49.

Sanders, A., 'On Mr. Darwin's Hypothesis of Pangenesis as Applied to the Faculty of Memory [Abstract and Comments]', *Journal of the Anthropological Society of London*, 8 (1870), cxlvi – clii.

Schacter, D. L., *Forgotten Ideas, Neglected Pioneers: Richard Semon and the Story of Memory* (Philadelphia – Hove: Psychology Press, 2001).

Schelling, F. D. J., *System of Transcendental Idealism (1800)*, trans. by P. Heath with an Introduction by M. Vater (Charlottesville: The University of Virginia, 1978).

Schulkin, J., ed., *Allostasis, Homeostasis, and the Costs of Adaptation* (Cambridge: Cambridge University Press, 2004).

Shaw, G. B., 'The New Life Reviewed', in *Pen, Portraits, and Reviews* (London: Constable and Company, 1932), pp. 52 – 71.

Singh, R., S. Kølvraa, and S. I. S. Rattan, 'Genetics of Longevity with Emphasis on the Relevance of HSP70 Genes', *Frontiers in Science*, 12 (2007), 4504 – 13.

Smith, J. A., and W. D. Ross, eds, *The Works of Aristotle*, 11 vols (Oxford: Clarendon Press, 1908 - 1952).

Stringer, C. B., and P. Andrews, 'Genetic and Fossil Evidence for the Origin of Modern Humans', *Science*, 295 (1988), 1263 - 68.

Stringer, C. B., and R. Grün, 'Palaeoanthropology. Time for the Last Neanderthals', *Nature*, 351 (1991), 701 - 02.

Taylor, J. B., and S. Shuttleworth, *Embodied Selves: An Anthology of Psychological Texts, 1830 - 1890* (Oxford: Clarendon Press, 1998), p. xiv.

Thangaraj, K., and others, 'Reconstructing the Origin of Andaman Islanders', *Science*, 308 (2005), 996.

Thorne, A. G., and M. H. Wolpoff, 'Regional Continuity in Australasian Pleistocene Hominid Evolution', *American Journal of Physical Anthropology*, 55 (1981), 337 - 49.

Thorne, A. G., and others, 'Australia's Oldest Human Remains: Age of the Lake Mungo 3 Skeleton', *Journal of Human Evolution*, 36 (1999), 591 - 612.

Torroni, A., and others, 'A Signal, from Human mtDNA, of Postglacial Recolonization in Europe', *American Journal of Human Genetics*, 69 (2001), 844 - 52.

Torroni, A., and others, 'Asian Affinities and Continental Radiation of the Four Founding Native American mtDNAs', *American Journal of Human Genetics*, 53 (1993), 563 - 90.

Torroni, A., and others, 'Classification of European mtDNAs from an Analysis of Three European Populations', *Genetics*, 144 (1996), 1835 - 50.

Torroni, A., and others, 'Harvesting the Fruit of the mtDNA Tree', *Trends in Genetics*, 22 (2006), 339 - 45.

Underhill, P. A., and T. Kivisild, 'Use of Y Chromosome and Mitochondrial DNA Population Structure in Tracing Human Migrations', *Annual Review of Genetics*, 41 (2007), 539 - 64.

Wais, K., 'The Symbiosis of the Arts', Address to the Second International Congress of Modern Literary History (Amsterdam: September, 1935), trans. by G. A. Richardson, *Yearbook of Comparative and General Literature*, 31 (1982), 78 - 95.

Warner, H., 'Longevity Genes: from Primitive Organisms to Humans', *Mech. Age. Dev.*, 126 (2005), 235 - 42.

Watson, E., and others, 'Mitochondrial Footprints of Human Expansions in Africa', *American Journal of Human Genetics*, 61 (1997), 691 - 704.

Willey, B., *Darwin and Butler: Two Versions of Evolution* (London: Chatto and Windus, 1960).

Williamson, G. S., *The Longing for Myth in Germany. Religion and Aesthetic Culture from Romanticism to Nietzsche* (Chicago and London: The University of Chicago Press, 2004).

Woolf, V., 'A Man with a View' (1916), in *Contemporary Writers* (London: Hogarth Press, 1965), pp. 28 - 35.

Yates, F. E., 'Order and Complexity in Dynamical Systems: Homeodynamics as a Generalized Mechanics for Biology', *Math. Comput. Model.*, 19 (1994), 49 - 74.

Part II: Originating Communication: Behaviours, Emotions, Languages

Abstracts

Augusto Vitale

The Evolutionary Origin of Human Behaviour: the Case of Altruism

In order to study the origin of human behaviours in an evolutionary perspective, different hypotheses about the impact of biological and cultural determinants on the development of human behaviour will be explored, and altruistic behaviour will be examined as a case study.

Whether altruistic behaviours of non-human animals reveal the origins of human altruistic behavioural patterns is a complex topic. Examples of parental care, mutualism and reciprocal altruism in non-human primates, our phylogenetically closest relatives, will be analysed in order to assess how the study of animal behaviour can improve our understanding of the origin of empathy and altruistic behaviours in humans.

The two major theories enunciated by Frans de Waal will be discussed: according to the "veneer theory", which regards humans as inherently selfish, ethics allow humans to keep the nastiness of the evolutionary process at bay. The other theory considers humans as naturally inclined towards cooperation and altruism. Non-human and human primates live in highly complex social systems, which function better when cooperation plays an important role in the relations among its members. Cooperation and altruism are engendered by 'empathy', namely the possibility to perceive the state of difficulty or negative condition which a conspecific faces in a particular moment.

The necessity to create cohesion between the members of a complex social system through cooperation and altruism, mediated by empathic capabilities, allows for explaining the origin of altruistic behaviours in terms of an evolutionary continuity between humans and non-humans. In other words, the existence of cooperative behaviours and altruism in our social ethogram could have a long evolutionary history.

Marco Dondi / Sergio Agnoli / Laura Franchin

A New Look at the Very Early Origins of Smiling

Of all the facial expressions of emotions, smiling has been studied empirically since the birth of psychology and has continued to be a central topic of this discipline until today. The reason for this long-standing interest probably derives from the fact that smiling appears to be one of the most complicated expressions in terms of the conditions that precipitate it, the functions it serves, and the developmental trajectories it depicts. Smiling is, in effect, an expressive behaviour that, during development, changes in morphology, in timing, in eliciting contexts, and in function. Its global facial configuration, produced by the contraction of the *zygomaticus major* muscles, must be regarded as an essential component of nonverbal social communication originating in the first few months of life, although its underlying 'emotions' or 'meanings' undergo extensive developmental and cultural transformations.

Smiling is observable as a stable pattern of facial movements present since birth, in term and pre-term neonates. Unlike 'social smiling', that emerges later (at about 2 months of age) and has been analysed in several early and more recent empirical contributions, little is known about the significance of neonatal smiles or their developmental relevance for later smiling. Neonatal smiling is considered a type of behaviour which occurs during sleep in the absence of recognized external or internal (visceral) stimuli. For this reason it is known in literature as *reflexive*, *spontaneous* or *endogenous* smiling.

The aim of this chapter is to present studies conducted since 2002: for the first time, the very early origins of smiling in pre-term and full-term neonates have been systematically investigated by using anatomically based facial coding techniques, such as FACS (2002) and Baby FACS (in press). In particular, contrary to what has been generally reported in literature, we found that Duchenne smiling (a form of smile associated in adults with positive emotional activation) is frequent in full-term and pre-term neonates, and that it is exhibited by infants in all behavioural states examined (even while neonates are awake). Unexpectedly, we also found a frequent and very brief motor activity of the *zygomaticus major* which has never been described in neonates, and an association between active (REM) sleep and long-lasting Duchenne smiles.

Laura Bafile

The Debate on the Origins of the Language Faculty in the Perspective of the Biolinguistic Programme

This chapter deals with the debate about the origins of language, a topic which, after more than a century of almost complete lack of interest, is now arousing conspicuous attention in linguistics as well as in other scientific fields. Interestingly, the attention devoted to origins is strictly connected with the conception of language which prevails in a certain period. Between the nineteenth and the twentieth century, the fundamental questions of mainstream linguistics did not address the philosophical problem of the nature of language and, consequently, the issue of origin was virtually ignored. In the last decade, owing to generative linguistics, the focus has shifted again towards language as a mental faculty of human beings. The question about the origin of language has thus been approached with renewed interest, and a lively debate has arisen.

Although the hottest dispute is between innatist and anti-innatist viewpoints, a variety of positions also characterises the biolinguistic framework, according to which language is considered a biological, species-specific property of human kind. In this perspective, the debate is between the adaptationist and the exaptationist views of language evolution. The former hypothesis focuses on the idea that the abilities which generate language have gradually evolved in our own species on the basis of an evolutionary design for communication. The latter hypothesis is founded on the idea that the origin of language is rather to be found in the emergence of recursivity, considered as the essential property of the language faculty, while each of the abilities that form such faculty in its broad sense may have had its own evolutionary history and may have appeared in humans, as well as in other animals, for reasons other than communication.

Keith Hoskin

Origins or Geneses? The Roles of Invention

While origins and geneses are both well-formed words, the latter is less used. It is employed here to suggest the importance of capturing the temporality of significant human breakthroughs and the extent to which historically as well as biologically they are dependent upon prior acts undertaken by the living but in form, action and consequences in no way predetermined by those prior acts. Additionally, it is argued that particularly significant moments of genesis may enable new kinds of human 'invention'.

Particular attention is given to the eighteenth-century epistemic shift identified in Michel Foucault's *The Order of Things* from a 'representational' *episteme* to the modern knowledge regime dominated by knowledge 'disciplines'. It is argued that the 'genesis' of this shift can be understood through a *genealogical* reinterpretation of the events that Foucault studied in *The Order of Things* using his archaeological method. On that basis a genealogy is then traced not only of the emergence of the modern knowledge disciplines, but of modern modes of discipline and self-discipline, including modern 'management'. It is suggested that management may be seen as an 'invention' which is only made possible following the genesis of the 'disciplinary' way of seeing, thinking and acting, drawing on a historical reinterpretation of Alfred Chandler's seminal history of management and modern business enterprise, *The Visible Hand*. In this way we may perhaps begin to think further on the complementarities and differences between origins, geneses and modes of invention.

Augusto Vitale

The Evolutionary Origin of Human Behaviour: the Case of Altruism

I. Introduction

How can we trace the origins of human behaviour? What is the most rational way to explain the evolution of human behaviour by placing species-specific traits among those exhibited by other species, in obvious absence of behavioural fossils? The answer must not come from a comparison between the human species and another particular species, but rather from the effort to see humans in the broader evolution of a mammalian pattern.

The aim of the present contribution is to discuss the origins of human behaviour within an evolutionary perspective. The question we want to address is: where do humans fit in a behavioural evolutionary scenario? Altruistic behaviour will be presented as a special case study, starting from the assumption that aspects of altruistic behaviours of non-human animals are useful to identify the origins of altruistic behavioural patterns which characterise our own species. The evolutionary perspective involves a comparative point of view: the differences and similarities in behaviours between different species are temporally and spatially contextualised in order to find a continuum of behavioural patterns across different taxonomic groups. However, because the focus of our study is the relationship between non-human and human behaviours, we will analyse different cases of altruistic behaviours mostly in species phylogenetically closer to humans, that is, the non-human primates. Indeed, numerous altruistic behaviours observed in non-human primates can be defined as examples of reciprocal altruism. By studying their sophisticated cognitive abilities we can gain a deeper insight into the study of the origin and manifestation of altruistic behaviours in humans.

Thus, primatology can be regarded not only as a field of enquiry into the similarities/differences between human and non-human, but also as a potentially productive meeting place for scientific and humanistic areas of research.

II. From Non-human to Human Behaviour

Konrad Lorenz, one of the founders of modern ethology, and winner of the Nobel Prize for Medicine in 1973, pointed out that knowledge obtained from the study of animal behaviour ought to be applied to gain a better understanding of human behaviour. This argument implies a behavioural continuity between non-human and human animals. What are the characteristics of such continuum? In other words, where does our behaviour come from?

Charles Darwin offered the inescapable conclusion that our behaviour, like other species, is a product of biological evolution. However, humans show a remarkable behavioural flexibility and complexity, characterised by elements of consciousness and abstraction of thoughts. Although in *On the Origin of Species* (1859) [1] Darwin never clearly mentioned the evolution of human behaviour, it was evident that he believed in the gradual evolution of our present forms (including behaviour) from non-human ancestors. However, a few years later, in *The Descent of Man and Selection in Relation to Sex* (1871), he affirmed that human behaviour is in some aspects like that of other animals, but also unique. Differences and similarities are the results of Natural Selection. [2]

Felicity Huntingford has indicated four main points of view in relation to the characterization of human behaviour and its evolution: [3]

(i) Human behaviour is determined by strong biological predispositions, where genes play a crucial role. These predispositions, which could be indicated as instincts, are shaped by Natural Selection, and have been inherited from our non-human ancestors. Two considerations derive from this point of view: a) learning processes produce variation of a biologically determined character and b) other animals, especially non-human primates, can provide information on the origins of human behaviour.
In the mid-Seventies, a sustained use of this perspective resulted in the development and success of an area of behavioural research, called Sociobiology, which was aimed at discovering genetic determinism in the field of human behaviour. [4] According to the reductionist framework, the nature of complex behavioural phenomena can be understood by reducing them to the interactions of their simpler parts; if its validity is accepted, it can be used to study our fear of snakes as well as our appreciation of art.

1 Charles Darwin, *On the Origin of Species* (London: John Murray, 1859).
2 Charles Darwin, *The Descent of Man and Selection in Relation to Sex* (London: John Murray, 1871).
3 Felicity A. Huntingford, *The Study of Animal Behaviour* (London: Chapman & Hall, 1984).
4 Edward O. Wilson, *Sociobiology: The New Synthesis* (Cambridge: Cambridge University Press, 1975); Ullica Segerstrale, *Defenders of the Truth: The Sociobiology Debate* (Oxford: Oxford University Press, 2001).

However, genetic predispositions can be influenced by the environment, therefore manifestations of genetically predisposed behavioural traits are not expected to be present in the whole ethogram of a particular species.

(ii) Culture is important in determining human behaviour. The products of the human brain can often be considered the result of Natural Selection. In this view, cultural behaviours which promote the inclusive fitness of an individual will be passed on to future generations, otherwise those behavioural traits will disappear in that particular population. Cultural modifications seem to invalidate the possibility of deriving direct information about human behaviours from non-human animals. However, human behavioural patterns are explained in terms of inclusive fitness, and then framed in a strong biological-evolutionary framework.

(iii) The cultural environment in which we develop is an important determinant of human behaviour, but it is also the result of the process of cultural evolution which, though very similar to Darwinian Natural Selection, is not one of its products. What distinguishes this point of view from the previous one is that cultural behaviours spread successfully within a population, but not in relation to their effect on the direct and/or inclusive fitness of the individuals showing that particular behaviour. In this case, biological evolution can be used as a model for understanding the origin and manifestations of human behaviour traits, but it does not provide direct explanations. In other words, culturally modified human behaviours are analogous, but not homologous, to non-human behaviours modified by Natural Selection.

(iv) Only culture determines human behaviour. Our culture exerts such a strong influence on what we do that it overrides everything else. Non-human animal behaviour can perhaps explain something about very simple human behaviours, but there is no use in comparing human to non-humans when we want to explain, from an evolutionary point of view, human social systems and man-made institutions. Evolutionary inheritance is taken into account only in the case of very broad behavioural predispositions. Culture, then, eliminates the effect of any genetic-related variation in behaviour, differently from what we see in non-human species.

Whatever point of view one chooses to apply, evidence from a variety of cross-cultural sources indicates that certain human behavioural traits found their origin in an inherited predisposition. For example, a series of facial expressions related to particular emotions appear to be rather widespread and consistent in different human populations, belonging to distant and different cultures. [5]

5 Irenaus Eibl-Eibesfeldt, *Human Ethology* (New York: Aldine de Gruyter, 1989).

Darwin himself, in *The Expression of the Emotions in Man and Animals* (1872), already noticed that many emotional expressions are universal and common among young and older people in different populations and cultures. [6]

III. Comparisons with Other Species

Which species should we compare our species with? To compare our behaviour with the behaviour shown by a particular selected species can result in a useless effort, because the answer depends very much on the animal we have chosen. It is more fruitful to focus on a more general picture, rather than search for similarities with a particular selected species. Therefore, how do humans fit within a general behavioural scenario shared by, for example, a large number of mammals?

With regards to human cognition there are two possible alternative methodological approaches.

Firstly, the cognitive capacities of phylogenetically distant species living in a similar environment could be analysed and compared. Similarities in such abilities would have then evolved independently from common ancestry (analogy approach).

Otherwise, the study of the cognitive capacities of species phylogenetically closer to humans could help us trace the evolutionary origin of such capacities, in relation to shared ancestry (homology approach). This approach focuses on phylogenetic constraints such as brain size and/or architecture, and monkeys and apes are the main subjects of such comparisons.

Both methodologies are part of the ethological approach to the study of the evolution of behaviour. [7] In order to analyse the evolution of altruism, elements from both approaches will be used.

IV. The Case of Altruistic Behaviour

The meerkat mongoose (*Suricatta suricata*) is a small carnivore living on the South African plains. It differs from most mongooses, which are usually solitary and live in family groups larger than single-family units. The females of this species organise communal nesting, that is, each adult female cares not only for

6 Charles Darwin, *The Expression of the Emotions in Man and Animals* (London: John Murray, 1872).

7 A. C. Kamil, 'On the Proper Definition of Cognitive Ethology', in *Animal Cognition in Nature*, ed. by R. P. Balda, I. M. Pepperberg, and A. C. Kamil (San Diego: Academic Press, 1998), pp. 1–28.

her own offspring, but also for the offspring of other females. The behaviour of the baby merkaats waiting at the nest for the females to deliver food is particularly interesting. Researchers from the University of Cambridge have discovered that if the offspring scream all together at the same time they will obtain, in average, more food *pro capite*, than if they scream in a less coordinate manner. [8] Therefore, it would seem that the baby merkaats are behaving cooperatively, because each one is helping the others to obtain more food. However, this behaviour could be interpreted in an alternative way: each baby merkaat tries to scream louder than the others, in order to obtain more food for itself. The effect is the same, but the motivation is rather different from an altruistic one.

Are baby merkaats an example of altruistic behaviour, or are they just behaving in a selfish manner? In more general terms, is the behaviour of this small sized carnivore telling us something about our own behaviour, and in particular on the origin of our altruistic behaviours? Because altruistic behaviour allows us to compare human and non-human species, it helps us understand particular behavioural traits of humans.

It is called altruistic any behaviour increasing the chances of survival and/or reproduction of the receiver, to the disadvantage of the altruist. Altruism poses an interesting theoretical problem, because it appears to be in contradiction with the Darwinian theory of Natural Selection. As a matter of fact, Darwin suggested that for each individual the characteristics, including behavioural ones, which improve the chances to survive and reproduce, are favoured characteristics (these chances are indicated as individual fitness). How is it then possible that a behaviour, which penalises the individual displaying it and favours other individuals, can be positively selected?

During the development of modern ethology brought about by post-war scientists involved in the study of animal behaviour, different hypotheses have been formulated to explain the appearance of altruistic behaviours in animal societies. The most famous and influential hypothesis, defined as kinship selection, has been proposed by the biologist William Hamilton, [9] who investigated altruistic behaviours directed toward relatives. The social insects, the *hymenoptera*, are a classic example. Hamilton searched for a gene-based theory which could explain the existence of sterile castes in these insects. He observed that an altruistic behaviour could evolve if the individuals toward whom those acts are directed have genes in common with the altruistic individuals. Therefore, kin-

8 Tim H. Clutton-Brock, P. N. Brotherton. M. J. O'Riain, A. S. Griffin, D. Gaynor, L. Sharpe, R. Kansky, Marta B. Manser, and G. M. McIlrath, 'Individual Contributions to Babysitting in a Cooperative Mongoose, *Suricata Suricatta*', *Proceedings of the Royal Society of London*, 267 (2000), 301–05.

9 W. D. Hamilton, 'The Genetical Evolution of Social Behaviour', *Journal of Theoretical Biology*, 7 (1964), 1–52.

ship selection will favour the reproduction, in a particular population, of those genes that influence the behaviours contributing to the survival and reproduction of related individuals. The fundamental intuition of Hamilton was to establish a relationship between the social systems of these invertebrates and reproductive mechanism and sex determination. Therefore, Hamilton considered that in the *hymenoptera* societies, particular individuals (the workers) do not reproduce in order to better help the reproduction of a fertile individual (the queen) and the care of her offspring. Furthermore, in these societies other individuals (the soldiers) sacrifice their lives to defend the nest. It is important to point out that the females develop from fertilised eggs, whereas the males come from parthenogenetic eggs: the females are diploid, whereas the males are haploid. These genetic profiles cause asymmetry in the coefficient of relatedness. Thus, twin sisters will have an average of 75 % of genes in common, whereas the relatedness between mother and offspring will be 50 %. Hamilton concluded that the behaviours of the workers aimed at producing sisters rather than daughters is positively selected, because a higher percentage of genes in common will then be passed to the next generation.

Perhaps, more interesting cases occur when help is offered to individuals who are not closely related: this is the case of mutualism, which occurs when a group of individuals cooperate to reach an aim, and every individual gains an immediate and clear advantage. Savannah hunters, such as lions (*Panthera leo*), have been observed to cooperate when hunting a dangerous and big prey, such as a water buffalo, or a particularly fast and elusive one, such as a gazelle. However, if the prey is too small and easy to catch, mutualism is avoided, because an individual will tend not to share a small prey: in this case the cost of sharing is higher than the cost of hunting by itself.

Another interesting case of altruism is called reciprocal altruism. It occurs when altruistic acts are shown between individuals belonging to the same social group, and reciprocity is diluted in time. Reciprocity in time makes it possible that altruistic traits, in average, are shown by all the members of a social group and, therefore, all the individuals can be both altruistic and benefit from the altruism of the others. Examples of this particular case of altruistic behaviour can be observed in non-human primates.

V. Altruism in Non-human Animals: the Case of Non-human Primates

Most non-human primate species generally behave like other mammals: the females tend to remain at home, whereas males migrate in search of new territories and social groups. These sex-related differences in social behaviour can influence the ways and time when to form alliances with members of the same sex. For example in Old World monkeys, such as in different species of the genus *Macaca*, females form alliances with members of the same matrilineal line, and this behaviour is facilitated by the tendency shown by the females not to leave the group. For example, a young female who belongs to a more dominant matrilineal line can win a contrast with a more adult female belonging to another line, because she can count on the help of other females of the same dominant line. On other occasions, females of lower social status can form alliances and chase away a more dominant female, causing a long lasting change in the dominance hierarchy of that particular group. [10] These behavioural phenomena can be identified as forms of mutualism, as illustrated above. Cooperating individuals perform mutualistic acts.

Non-genetically related males can form temporary alliances as well. An example comes from the alliance formed by two males of olive baboons (*Papio anubis*), to divert another male from guarding a female in oestrus. In this species, when a female becomes sexually receptive, she forms a strong bond with the dominant male. However, other males are present in the group, for whom it is more difficult to reproduce. Therefore, these males can try to form alliances in order to steal, at least temporarily, the female from the attention of the dominant male. In this case we can observe a typical altruistic behaviour: one of the two males forming the alliance attacks the dominant male, diverting him from the desired female; the second male takes advantage of the situation and mates with the female temporarily left alone. The attacking male of the alliance gives up the possibility of mating with the female and possibly increasing his fitness and, at the same time, favours the reproductive chances of his ally. [11] We can notice an important difference from the cases of mutualism illustrated before. One of the members of the alliance does not only appear to gain no advantage, but he also risks life when attacking the dominant male in order to benefit the ally. However, this is a case of reciprocal altruism, which appears to be significantly present in the social ethogram of non-human primates.

---

10 P. L. Vasey, 'Interventions and alliance formation between female Japanese macaques, *Macaca fuscata*, during homosexual consortships', *Animal Behaviour*, 52 (1996), 539–51.

11 D. L. Cheney, 'The acquisition of rank and the development of reciprocal alliances among free-ranging immature baboons', *Behavioural Ecology & Sociobiology*, 2 (1977), 303–18.

In order to consider this situation as a case of reciprocal altruism, the roles of the two allies must necessarily reverse at some point in time so that, in average, each one has his own possibility of mating and increasing his own fitness. A condition under which reciprocal altruism can effectively work is the ability to spot the cheaters and avoid them: the cheaters will not reciprocate and this is a crucial limitation. Whether monkeys are able to identify such individuals is still very much debated.

Researchers are discussing what type of reciprocity can be expected from different species, in relation also to their cognitive abilities. Brosnan and de Waal, in particular, have identified three instances of reciprocity. The least complex form of reciprocity, in cognitive terms, is based on dyadic relationships (special bonds between two individuals, kinship). A more complex form is based on the affinity of social attitudes between the individuals involved: this form has been observed in capuchin monkeys and chimpanzees. Thirdly, only humans, and perhaps chimpanzees, show what can be called calculated reciprocity, which requires a sort of mental scorekeeping. [12]

In reciprocal altruism the altruistic individual expects to be, sooner or later, rewarded. However, in some cases altruistic acts in non-human primates do not appear to entail an exchange of favours. Data recently collected on a captive colony of common marmosets (*Callithrix jacchus*) suggest that monkeys perform altruistic acts, denominated 'prosocial', which benefit the members of their own social group but not them, apparently: the active offer of food to social mates is one of these acts. In this particular experiment food was offered to social mates who were neither reciprocating, nor had genetic relationships. The performance of food offering behaviour provides data which correlate positively with the time two particular individuals have been living together, the number of offspring, and the frequency of other prosocial behaviours. Researchers were also able to discard the hypothesis that reciprocal altruism could explain the observed behaviour. They suggest that prosocial attitudes may originate from cooperative breeding, because marmosets and humans are the only primates who are both cooperative breeders and show such unsolicited prosociality, which very often manifests itself in altruistic acts. [13]

Obviously, among non-human primates, chimpanzees (*Pan troglodytes*) are favourite species of study in the area of altruistic behaviours. Warneken and colleagues studied altruism in a semi-captive colony of chimpanzees. These researchers were able to show that chimpanzees were willing to help unfamiliar

12 Sarah F. Brosnan and Frans B. M. de Waal, 'A Proximate Perspective on Reciprocal Altruism', *Human Nature*, 13 (2002), 129–52.

13 Judith M. Burkart, Ernst Fehr, Charles Efferson, and Carel P. van Schaik, 'Other-regarding preferences in a non-human primate: Common marmosets provision food altruistically', *Proceedings National Academy Science*, 104 (2007), 19762–66.

humans to reach a stick, with no apparent reward. [14] One of the possible interpretations of the results was that these animals, used in different experimental protocols, have learned that pleasing a human would, sooner or later, result in a reward. So, it follows that these chimpanzees would act altruistically owing to the long-lasting experimental way of living, rather than to true altruistic propensity. Such reciprocal altruism, which occurs between chimps and human experimenters, is thus not spontaneous prosocial altruism. However, in trying to overcome such criticism, Warneken and his team performed a further experiment, where chimpanzees were asked to choose whether or not to help an unrelated conspecific to unchain a door. The results suggested the possibility that these apes show true altruistic behaviour. The help was offered even when it required some effort, and no obvious expectation of a reward was observed. Adding the fact that recipients of altruistic behaviours were unrelated individuals, the behaviour of these animals was remarkably similar to altruism and cooperative behaviours shown by the human species.

Warneken's observations do not match with the results obtained in other laboratories. In other experiments, on the one hand chimpanzees sometimes rejected exchanges in which they felt penalised compared to others, which may suggest the presence of a sort of 'sense of fairness'; on the other hand, there was no evidence that they were averse to interactions in which they benefited more than others. [15]

These contradictory data suggest that the manifestation of altruistic behaviours in this primate species, but perhaps not only in chimpanzees, depends on the type of experimental protocol, individual differences, context and history of that particular group of animals. For example: how does the experimental history of a particular colony of captive primates affect their behaviour, in relation to the expectation of obtaining care and rewards?

Data coming from natural populations are necessary to obtain a better picture of this behavioural phenomenon. Frans de Waal affirms that animals in captivity are more likely to show spontaneous altruistic behaviours because they are less worried about dangers and necessities that influence their behaviour in the wild. We could also add that in captivity, as mentioned before, animals have a strong opportunity to learn the association between a certain act and the consequent reward. What calls for closer scrutiny is the occurrence of altruistic behaviours in the environmental context where the behavioural repertoire of a particular species has evolved.

14 Felix Warneken, Brian Hare, Alicia P. Melis, Daniel Hanus, and Michael Tomasello, 'Spontaneous altruism by chimpanzees and young children', *PloS Biology*, 5 (2007), e184.

15 Joan B. Silk, Sarah F. Brosnan, Jennifer Vonk, Joseph Enrich, Daniel J. Povinelli, Amanda S. Richardson, Susan P. Lambeth, Jenny Mascaro, and Steven J. Schapiro, 'Chimpanzees are indifferent to the welfare of unrelated group members', *Nature*, 437 (2005), 1357–59.

VI. From Non-human Primates to Human Primates

What does research on non-human species tell us about the origin and evolution of altruistic behaviours in humans? We have seen that, under certain circumstances, non-human primates show altruistic behaviours and different forms of cooperation. However, the consistency of cooperation in humans is overwhelming, compared to non-humans. Human tendency to cooperate seems to appear in infancy, as demonstrated by experiments in which 18-month-old children were observed to aid, not solicited, an adult in need of help to perform a task. [16]

Can we identify a common origin of altruistic behaviours shown by different species of primates (including our species) that would allow hypothesising a strong evolutionary adaptiveness? Are these characters homologous, that is, are they sharing a common evolutionary history?

Frans de Waal has suggested that humans are naturally inclined to develop cooperation and altruism. The reason beyond this idea is that cooperation is considered an essential feature of a functioning complex social system. Non-human and human primates are both characterised by complex sociality, which manifests itself in a web of diversified relationships. These systems can better function when cooperation plays an important role in the relations among its members. De Waal argues that the mechanism which enables the onset of cooperation is empathy, that is the possibility of perceiving the state of difficulty or negative condition(s) in which non-human or human individuals find themselves in a particular moment. [17] Empathy is generally attributed to humans alone, with the possible exception of non-human primates. However, researches have recently demonstrated that a variation in pain sensitivity in mice can be produced by exposure to their cage-mates in pain, but not to strangers. [18]

The need to keep together the members of a complex social system through cooperation, mediated by empathic abilities, is a strong argument in favour of an evolutionary continuity between humans and non-humans. This argument can be applied to both human and non-human complex social systems, and thanks to this idea we can suggest that the existence of cooperative behaviours in our social ethogram could have a long evolutionary history. The issue, which will not be dealt with here, is whether these behaviours can be regarded as the founda-

16 Felix Warneken and Michael Tomasello, 'Altruistic Helping in Human Infants and Young Chimpanzees', *Science*, 311 (2006), 1301–03.

17 Frans B. M. de Waal, *Primates and Philosophers* (Princenton: Princeton University Press, 2006).

18 Dale J. Langford, Sara E. Crager, Zarrar Shehzad, Shad B. Smith, Susana G. Sotocinal, Jeremy S. Levenstadt, Mona Lisa Chanda, Daniel J. Levitin, and Jeffrey S. Mogil, 'Social modulation of pain as evidence for empathy in mice', *Science*, 312 (2006), 1967–70.

tions for the evolution of morality in humans. Obviously, this argument requires collaboration between biologists and humanities scholars.

However, other researchers disagree with de Waal. For example, the American primatologist William McGrew argues that scientific evidence from the behaviour of non-human primates is not convincing enough to sustain that precursors of human moral behaviour are found in non-human primates. More data are needed in order to affirm that empathy, sympathy, consolation and gratitude are features that can be found in the social ethogram of non-humans. [19]

Although McGrew does not give much credit to it, the 'Veener Theory' is based on the idea that humans are inherently selfish and ethics helps them keep the nastiness of the evolutionary process at bay. In his Romanes lecture 'Evolution and Ethics' (1893) Thomas Henry Huxley, the fiercest defender of Darwin's theory, argued that Natural Selection is merciless, because it is generated by a deadly competition for scarce resources. Indeed, what is good for one individual is necessarily bad for another. Empathy and altruism thus appear to be unthinkable. In humans, ethics help to control selfish tendencies.

Therefore, while de Waal erects an evolutionary bridge between non-human and human altruistic behaviours, Huxley and his contemporary followers highlight a fracture between the behaviour of humans and other animals.

VII. Conclusions

The study of altruism in animals can be a powerful tool for understanding the origin of a particular aspect of human behaviour.

Cooperation and reciprocal altruism are essential parts of human (and humanitarian) behaviour but, as we have seen, they can be found in other animals as well, above all in non-human primates. Such an approach is in accord with an evolutionary framework characterised by a continuity of forms and behaviours between humans and other animals. Biological predispositions and culture play an important role in shaping our behaviours towards our conspecifics.

The recognition of common aspects of altruistic behaviour between human and non-human animals should not be taken as a shortcut to affirm the existence (or better, reification) of behavioural patterns guided by a sense of morality among non-human animals. Morality can be discussed in biological and adaptive terms, but cooperation between different areas of enquiry is needed to gain a more comprehensive picture. Because the study of origins of altruism

19 W. C. McGrew, Review of 'Primates and Philosophers. How Morality Evolved' by Frans de Waal, *PaleoAnthropology* (2007), 34–35.

poses questions on the origins of human morality, it requires contributions from both the sciences and the humanities.

A word of caution should be said about the point of view of the researcher who interprets data regarding the cooperative or non-cooperative nature of non-human behaviour. Are the baby merkaats behaving altruistically, or just acting selfishly? The case of altruism is an example in which the idea of continuity between non-human and human behaviour is subject to the influence of the researcher's personal values and critical parameters. Biology offers narrations of the facts of life, rather than absolute truths.

Marco Dondi / Sergio Agnoli / Laura Franchin

A New Look at the Very Early Origins of Smiling [1]

I. Introduction

Over the past forty years, a dramatic explosion of the interest in emotions has become evident throughout a wide variety of disciplines, including psychology, cognitive science, biology, sociology, anthropology, philosophy, neuroscience, psychiatry, and economics. [2] Science has got to understand that emotion is not a luxury. It plays a critical role in virtually all aspects of learning, reasoning, and creativity. [3] This new interest was reflected in the foundation of the International Society for Research on Emotions (ISRE) in 1984, a society where researchers from various disciplines and cultures come together to discuss issues of mutual interest with the emotions as central focus.

The multi-disciplinary approach to the study of emotion produced an exponential increase of empirical research. An impressive multitude of levels at which emotions can be studied, described, and analysed, emerged from this scenario. An attempt to summarize the central themes around which the current state of emotion research is organized was made at a symposium held in 2001. 'Feelings and Emotions: The Amsterdam Symposium' aimed at reviewing the major research domains in the study of emotions from a multi-disciplinary perspective.

One of the most challenging themes inspiring the Amsterdam Symposium concerned the nature and the origins of emotions. New insights and methodologies are currently available for the understanding of the evolutionary meaning of emotional phenomena. The recent debate is shifting from the

---

1 The authors wish to thank the infants and parents who participated in these studies. Sincere appreciation is also conferred to Dorinda Mosedale for thoughtful comments on an earlier draft of this chapter.

2 A. S. R. Manstead, N. H. Frijda, and Agneta Fischer, *Feelings and Emotions* (Cambridge: Cambridge University Press, 2004).

3 A. R. Damasio, 'Emotions and Feelings: A Neurobiological Perspective', in *Feelings and Emotions*, pp. 49 – 57.

question as to whether emotions are biological or cultural, to that of the extent to which biological givens and cultural influences determine emotions. [4] The old controversy between a biological and cultural view is being abandoned and replaced by the idea that emotions are both biologically and culturally determined. Recent theorizing presents an emergence of views that acknowledge the potential compatibility of biological, socio-cultural and psychological approaches to emotion. In particular, the study of the facial expressions involved in emotions is considered as a primary research domain for the understanding of the influences exerted by biological and socio-cultural factors in the emergence of emotions. There is a cross-cultural and universal affinity between specific facial expressions and particular categories of emotion. [5] The way this affinity establishes and/or emerges during ontogenesis is a central research area of the psychology of emotion.

The analysis of the changes in morphology, social display, and communicative function of the emotional facial expressions throughout development produced new insights into the study of the nature and functions of emotions. To outline the developmental trajectories of the emotional facial expressions has represented in psychology a widespread approach to the study of the meaning and origins of emotions. We have chosen to investigate smiling and its very early developmental trajectory.

II. From Duchenne and Darwin to Current Anatomically Based Facial Coding Systems

Since the publication of the *Mécanisme de la Physionomie Humaine* by Guillame Duchenne de Boulogne [6] in 1862, and of *The Expression of Emotions in Man and Animals* by Charles Darwin [7] in 1872, the behaviour involved in emotional response, that is facial expressions, postures, and gestures, has been considered a central issue to modern scientific inquiry in biology as well as in psychology. In particular, the publication of these books can be considered the starting point of the modern study of emotion and facial behaviour. The main assumption of Duchenne's work was that '[...] the laws that govern the expressions of the

4 *Feelings and Emotions.*

5 N. H. Frijda and Anna Tcherkassof, 'Facial Expressions as Modes of Action Readiness', in *The Psychology of Facial Expression*, ed. by J. A. Russell and J. M. Fernández-Dols (Cambridge: Cambridge University Press, 1997), pp. 78–102.

6 G. B. Duchenne (de Boulogne), *Mécanisme de la Physionomie Humaine ou Analyse électro-physiologique de l'Expression des Passions* (Paris: Jules Renouard Libraire, 1862).

7 Charles Darwin, *The Expression of Emotions in Man and Animals* (Chicago: University of Chicago Press, 1872, repr. 1965).

human face can be discovered by studying muscle action'. [8] The fact that emotional expressions in humans and animals arise by the same evolutionary mechanisms was first proposed by Darwin. He also assumed that there was a strong relationship between emotional states and related expressions both in humans and animals.

The accurate description of facial behaviour made by Duchenne was mainly enabled by the improvements in photography at the end of the 19th century, which permitted for the first time in history the freezing of facial actions and expressions. He used a new technique of electrical stimulation combined with photography, identifying and describing isolated facial muscular contractions with extreme accuracy. For most of his pictures Duchenne employed the photographer Adrien Tournachon, who was the brother of Felix Nadar, one of the best known early practitioners of photography in France. In *The Expression of Emotions* Darwin emphasized his debt to Duchenne by quoting his theories and by using several of his illustrations.

As Duchenne and Darwin took advantage of the improvements within technology into studying what had been defined as too fleeting and short lived to be investigated, more recent emotion theorists, such as Ekman and Izard, have taken advantage of the availability of analogical and digital video-recording in analysing the dynamics of facial behaviour. At the end of the Seventies a series of new methodologies for coding facial actions were built in the United States. [9] For all these modern coding systems the point of departure was the anatomical atlas [10] of the facial musculature, and the first task was an accurate description of the changes in facial appearance caused by different muscular contractions.

Today the most commonly used methodology in facial coding is the *Facial Action Coding System* (FACS) [11] by Paul Ekman and Wallace Friesen. In the construction of this coding methodology the authors explored and analysed nearly 10,000 combinations of facial muscle movements. Using this information they were able to create a measurement system that allowed for decomposing any observed facial expression into the muscular actions that produced it. The last

8 G. B. Duchenne de Boulogne, *The Mechanism of Human Facial Expression*, ed. by R. A. Cuthbertson (Cambridge: Cambridge University Press, 1990), p. 1.

9 Paul Ekman and W. V. Friesen, *Facial Action Coding System (FACS)* (Palo Alto: Consulting Psychologist Press, 1978); C. E. Izard, *The Maximally Discriminative Facial Movement Coding System (MAX)* (Newark: University of Delaware, Instructional Resources Center, 1979); C. E. Izard, L. M. Dougherty, and E. A. Hembree, *A System for Identifying Affect Expressions by Holistic Judgments (AFFEX)* (Newark: Instructional Resources Center, University of Delaware, 1983).

10 C. H. Hjorstijo, *Man's Face and Mimic Language* (Lund: Student-Literature, 1970).

11 *Facial Action Coding System (FACS)*; Paul Ekman, W. V. Friesen, and J. C. Hager, *Facial Action Coding System (FACS)* (Salt Lake City: Research Nexus division of Network Information Research Corporation, 2002).

version of FACS, published in 2002, identifies 9 basic facial actions (named Action Units, AUs) in the upper face and 18 in the lower face, as well as five levels of muscular intensity of contraction. Furthermore, FACS specifies 11 head positions and movements, 4 miscellaneous Action Units, 9 action descriptors (movements for which the anatomical basis is unspecified), 9 gross behaviours, 9 eyes positions and movements plus 4 visibility and supplementary codes. Taking into account several significant differences between adults and infants, Oster recently proposed a version of FACS re-adapted for infant face morphology: the *Facial Action Coding System for Infants and Young Children* (Baby FACS). [12] Two anatomically-based coding methods dedicated to infant facial emotional expressions were also built by Izard and co-workers: the *Maximally Discriminative Facial Movement Coding System* (MAX) [13] and *A System for Identifying Affect Expressions by Holistic Judgments* (AFFEX). [14]

Throughout the use of these methods, trained coders may distinguish for each of the facial actions the duration and its intensity by analysing extensively and in slow motion selected video-segments on a video-recorder or on a personal computer. For example, in finding the onset time of a facial action, coders proceed until the movement peaks and is clearly visible. They then have to move backwards until the movement stops and note the onset time, confirming or adjusting it by moving the tape forward and backwards across the estimated onset point. The offset time is obtained exactly in the same manner. This coding procedure based on a frame by frame video analysis is known in psychology as behavioural micro-analysis. The extensive use of these methods and of micro-analysis gave rise to hundreds of scientific reports on different emotional expressions, improving the quality of empirical data in the area of emotion research.

The main aim of this chapter is to present recent studies in which for the first time the very early origins of smiling have been investigated by using these anatomically based coding techniques. This analysis is preceded by a brief review about smiling and its early development and by an introductory paragraph on neonatal smiling.

---

12 Harriet Oster, *Baby FACS: Facial Action Coding System for Infants and Young Children* (Draft version revised April 2007, in press).

13 *The Maximally Discriminative Facial Movement Coding System (MAX).*

14 *A System for Identifying Affect Expressions by Holistic Judgments (AFFEX).*

III. Functions, Forms and Early Development of Smiling

Smiling is the most cross-culturally recognized human facial expression of emotion in Western as well as in Preliterate societies indicating, universally, a positive emotional state or activation. [15] Smiling is also universally considered as an expression of positive attitude towards cooperative and/or affiliative relationships with others. [16] The smile has in fact the capacity to operate as a signal of readiness for friendly interaction. This suggests that smiling is a biologically based and evolved behaviour that serves crucial adaptive functions in humans.

The adult smile also has social regulatory functions masking negative emotions such as displeasure, embarrassment, and anxiety. Besides expressing pleasure, joy, or happiness, it is used to signal fear, aggression, doubt, contempt, or defiance. This indicates that smiling is deeply influenced by cultural factors, which determine different uses and *display rules* [17] employed within different societies and contexts. In fact, cross-cultural research has clearly demonstrated that each human culture as well as each social group dictates through unwritten rules when, why and where it is convenient to smile. Given its complexity and its frequent use in social interaction, smiling has been studied since the birth of modern psychology, [18] continuing to be debated even today. [19]

Smiling facial configuration derives from the contraction of *zygomaticus major* muscles which enlarge and pull up the corners of the mouth. The recent anatomically based methods used for coding facial actions have described in great detail the appearance changes caused by the *zygomaticus major* activity, giving an objective and reliable account of its characteristics. For example, FACS

15 Paul Ekman, 'Universal and Cultural Differences in Facial Expressions of Emotions', in *Nebraska Symposium on Motivation 1971*, ed. by James K. Cole (Lincoln: University of Nebraska Press, 1972), pp. 207–83; Paul Ekman and W. V. Friesen, 'Constants across Cultures in the Face and Emotion', *Journal of Personality and Social Psychology*, 17 (1971), 124–29; J. A. Russell, 'Is there Universal Recognition of Emotion from Facial Expression? A Review of the Cross-cultural Studies', *Psychological Bulletin*, 115, 1 (1994), 102–41.

16 A. J. Fridlund, 'The Behavioral Ecology View of Smiling and other Facial Expressions', in *An Empirical Reflection on the Smile*, ed. by Millicent H. Abel (New York: Edwin Mellen Press, 2002), pp. 103–29.

17 'Universal and Cultural Differences in Facial Expressions of Emotion'; Paul Ekman and W. V. Friesen, *Unmasking the Face. A Guide to Recognizing Emotions from Facial Clues* (Englewood Cliffs: Prentice-Hall, 1975).

18 *The Expression of Emotions in Man and Animals*; *Mechanisme de la Physionomie Humaine*; Georges Dumas, *Le Sourire. Psychologie et Physiologie* (Paris: Presses Universitaires de France, 1948).

19 *An Empirical Reflection on the Smile*; D. S. Messinger and Alan Fogel, 'The Interactive Development of Social Smiling', in *Advances in Child Development and Behavior*, vol. 35, ed. by Robert Kail (Oxford: Elsevier, 2007), pp. 327–66; Erika Rosenberg and Paul Ekman, *What the Face Reveals: Basic and Applied Studies on Spontaneous Expressions Using the Facial Action Coding System (FACS)*, 2nd edn (New York: Oxford University Press, 2005).

and Baby FACS distinguishes amongst five different levels of muscular intensity of the *zygomaticus major* contraction (named AU 12, 'lip corner puller') and provides extensive information to differentiate zygomaticus motor activity from that produced by other facial muscles.

The basic facial appearance caused by the *zygomaticus major* contraction (named simple smile) may be enriched by the co-occurrence of other facial actions distinguishing, morphologically, at least 18 different types of smiles. [20] These, in some cases, have been suggested as displaying different meanings or states of consciousness in adults [21] as well as in infants. [22] Among them, the most studied smile configuration is the Duchenne smile, which consists, following FACS, in the co-occurrence of the 'lip corner puller' (AU 12) produced by the *zygomaticus major* muscles and the 'Cheek Raiser and Lid Compressor' (AU 6) produced by the *orbicularis oculi pars lateralis* muscles. This particular smile is known as Duchenne smile, because Duchenne de Boulogne [23] first brought attention to the anatomical concomitants of a genuine smile of enjoyment. Specifically, he observed that genuine smiles are associated with the contraction of the *orbicularis oculi*. By contrasts, fake non-enjoyment smiles generated by electrical stimulation consisted of the *zygomaticus major* contraction only. Today the Duchenne smile, on the basis of a consistent body of research which followed Duchenne's intuitions, is generally considered the prototypical expression of the emotion of joy. On the contrary, simple or non-Duchenne smiles are maintained to be social signals or unfelt smiles and are not regarded as expressions of a genuine emotion (see note 20 above).

Between the neonatal period and the second year of life, smiles develop from an almost entirely endogenous behaviour, which occurs in the absence of recognized external or internal stimuli, to intentionally shared positive messages. [24]

20 Paul Ekman, *Telling Lies: Clues to Deceit in the Marketplace, Politics, and Marriage* (New York: W. W. Norton, 1985); Paul Ekman, R. J. Davidson, and W. V. Friesen, 'The Duchenne Smile: Emotional Expression and Brain Physiology II', *Journal of Personality and Social Psychology*, 58, 2 (1990), 342–53.

21 'The Duchenne Smile'; Paul Ekman and W. V. Friesen, 'Felt, False, and Miserable Smiles', *Journal of Nonverbal Behavior*, 6 (1982), 238–52; Mark Frank, Paul Ekman, and W. V. Friesen, 'Behavioral Markers and Recognizability of the Smile of Enjoyment', *Journal of Personality and Social Psychology*, 64 (1993), 83–93.

22 Alan Fogel, H. C. Hsu, A. F. Shapiro, G. C. Nelson-Goens, and Cory Secrist, 'Effect of Normal and Perturbed Social Play on the Duration and Amplitude of Different Types of Infant Smiles', *Developmental Psychology*, 42, 3 (2006), 459–73; Alan Fogel, G. C. Nelson-Goens, H. C. Hsu, and A. F. Shapiro, 'Do Different Infant Smiles Reflect Different Positive Emotions?', *Social Development*, 9 (2000), 497–520; N. A. Fox and R. J. Davidson, 'Patterns of Brain Electrical Activity during Facial Signs of Emotion in 10 Month Old Infants', *Developmental Psychology*, 24, 2 (1988), 230–36.

23 *Mechanisme de la Physionomie Humaine.*

24 Laura Bolzani, D. S. Messinger, M. E. Yale, and Marco Dondi, 'Smiling in Infancy', in *An*

Throughout the first month smiles are most likely to happen when infants are drowsy or asleep. Between the first and the second month infants become more reactive to their environment and smile towards visual stimuli such as the human face or its schematic representation. During this period smiles are also occasionally elicited by the human voice and by other high-pitched sounds. Shortly after two month of age *social smiling*, [25] that is the smile systematically given to the human face, develops. From this point on smiling becomes a fundamental expression of emotion and an essential component of nonverbal social communication.

Both the frequency and the amount of time the infants spend smiling during interactions increase through to six months of age. [26] Among infants between one and six months of age, Duchenne smiling is more prevalent than non-Duchenne smiling when infants are gazing at their mothers and when their mothers are also smiling. [27] With development, more generally, infants become less passive and more active participants in the exchanges that elicit smiles. [28] Among 10-month-old infants, Duchenne smiles are associated with mother's smiling approach, while non-Duchenne smiles are related with the approach of an impassive stranger. [29] By using EEG recording, Fox and Davidson found that at this age Duchenne smiles are associated with left frontal activation, the pattern of cerebral activity repeatedly found with positive affect in adults. [30]

Empirical Reflection on the Smile, pp. 111–35; R. N. Emde and R. J. Harmon, 'Endogenous and Exogenous Smiling Systems in Early Infancy', *Journal of the American Academy of Child Psychiatry*, 11 (1972), 177–200; L. A. Sroufe, *Emotional Development* (New York: Cambridge University Press, 1996); P. H. Wolff, *The Development of Behavioral States and the Expression of Emotions in Early Infancy: New Proposals for Investigation* (Chicago: University of Chicago Press, 1987).

25 R. A. Spitz, *The First Year of Life: A Psychoanalytic Study of Normal and Deviant Development of Object Relations* (New York: Basic Book, 1965); R. A. Spitz and K. M. Wolf, 'The Smiling Response: A Contribution to the Ontogenesis of Social Relations', *Genetic Psychology Monographs*, 34 (1946), 57–125.

26 'Smiling in Infancy'.

27 D. S. Messinger, Alan Fogel, and K. L. Dickson, 'What's in a Smile?', *Developmental Psychology*, 35 (1999), 701–08; D. S. Messinger, Alan Fogel, and K. L. Dickson, 'All Smiles are Positive, but Some Smiles are More Positive than Others', *Developmental Psychology*, 37, 5 (2001), 642–53.

28 'Smiling in Infancy'.

29 'Patterns of Brain Electrical Activity'.

30 'The Duchenne Smile'.

IV. Neonatal Smiling

Neonatal smiling is the first opportunity to examine this facial display in human development. Smiling, in fact, can be observed as a stable pattern of facial movements present at birth, in term [31] and pre-term neonates. [32] Unlike the *social smiling* that has been analysed in several empirical contributions, very little is known about the significance, form, and recognisability of neonatal smiles or their developmental relevance for later smiling. In particular, like other early spontaneous motor patterns exhibited during neonatal sleep, such as startles, mouthing, reflexive sucks, and also penile erections in males, neonatal smiling is generally considered to be a behaviour which occurs in the absence of recognized external or internal stimuli. For this reason, it is known in literature as *reflexive, spontaneous* or *endogenous* smiling. [33]

Anecdotal and early observational evidence have suggested for many years that newborn smiling is exclusively associated with the state of active sleep (irregular, paradoxical or REM sleep) and drowsiness (the transition between alertness and REM sleep and vice-versa). In fact, during research done in the Sixties neonatal smiling was found to occur almost exclusively during REM states, therefore indicating that endogenous smiling is a consistent correlate of this behavioural state. In support of this claim, it was also reported that neonatal spontaneous smiles never or very rarely occur while the infant is fully awake or deeply asleep. For these reasons, REM endogenous smiling has been considered one of the most well circumscribed state-related behaviours found in the neonatal period. [34]

31 R. N. Emde and K. L. Koenig, 'Neonatal Smiling and Rapid Eye Movement States', *Journal of the American Academy of Child Psychiatry*, 8 (1969), 57 – 67; R. N. Emde and K. L. Koenig, 'Neonatal Smiling, Frowning, and Rapid Eye Movement States. II Sleep-Cycle Study', *Journal of the American Academy of Child Psychiatry*, 8 (1969), 637 – 56; A. F. Korner, 'Neonatal Startles, Smiles, Erections, and Reflex Sucks as Related to State, Sex, and Individuality', *Child Development*, 40 (1969), 1039 – 53; P. H. Wolff, 'Observations on Early Development of Smiling', in *Determinants of Infant Behavior*, ed. by B. M. Foss, vol. 2 (New York: John Wiley, 1963), pp. 113 – 38.

32 R. N. Emde, R. D. Mc Cartney, and R. J. Harmon, 'Neonatal Smiling in REM States, IV: Premature Study', *Child Development*, 42 (1971), 1657 – 61; *The Development of Behavioral States.*

33 'Endogenous and Exogenous Smiling Systems in Early Infancy'; 'Neonatal Startles, Smiles, Erections, and Reflex Sucks as Related to State, Sex, and Individuality'; L. A. Sroufe and Everett Waters, 'The Ontogenesis of Smiling and Laughter: A Perspective on the Organization of Development in Infancy', *Psychological Review*, 83, 3 (1976), 173 – 89; *The Development of Behavioral States.*

34 'Neonatal Smiling and Rapid Eye Movement States'; 'Neonatal Smiling, Frowning, and Rapid Eye Movement States. II Sleep-Cycle Study'; 'Neonatal Startles, Smiles, Erections, and Reflex Sucks as Related to State, Sex, and Individuality'; P. H. Wolff, 'Observations on Newborn Infants', *Psychosomatic Medicine*, 21, 2 (1959), 110 – 18; 'Observations on Early Development

Based on the studies performed during the Sixties, neonatal smiling has been generally defined as the bilateral sideward and upward drawing of the corners of the mouth into a partial U-shaped configuration that is achieved by the isolated contraction of the *zygomaticus major* muscles. The neonatal smile was in fact typically described at birth without reference to the action of the circular muscle surrounding the eyes, that is the action of the *orbicularis oculi*. In particular, it has often been suggested that smiling involving this muscle (which we define as the Duchenne smile) is a more mature form of smiling that is not present in neonates and appears at some later developmental point with the emergence of *social smiling*. [35] The assumption that Duchenne smiles were absent at birth was mainly and indirectly reinforced by the tradition of Duchenne smile research in adults. As mentioned before, according to this tradition Duchenne smiling is considered a prototypical expression of joy and is thought to be strictly associated with enjoyment. This overall data supported the inference that the neonatal smile, which is a simple smile, was non-emotional, while the Duchenne smile, exhibited later in development by an alert infant, was a true emotional expression.

Neonatal smiling is not generally considered to be a true, genuine or mature expression of emotion, but just a precursor. [36] This hypothesis also implies the widespread assumption that the emergence of *social smiling* (around the second month), may be the marker of the established relationship or coordination between the neural components regarding a positively tuned internal state and the corresponding facial expression. [37]

Occasionally, endogenous smiling has also been grouped amongst primitive behaviours and reflexes which may reappear in adulthood in patients showing severely impaired cortical function and little voluntary movement due to severe brain damage. This suggests that endogenous smiling is mainly controlled by the subcortical brain region.

---

of Smiling'; P. H. Wolff, *The Causes, Controls and Organization of Behavior in the Neonate* (New York: International Universities Press, 1966).

35 'Endogenous and Exogenous Smiling Systems in Early Infancy'; Alan Fogel and Ester Thelen, 'Development of Early Expressive and Communicative Action: Reinterpreting the Evidence from a Dynamic Systems Perspective', *Developmental Psychology*, 23, 6 (1987), 747–61; 'Neonatal Startles, Smiles, Erections, and Reflex Sucks as Related to State, Sex, and Individuality'; D. S. Messinger, Alan Fogel, and K. L. Dickson, 'A Dynamic System Approach to Infant Facial Action', in *The Psychology of Facial Expression*, pp. 205–26; Harriet Oster and Paul Ekman, 'Facial Behavior in Child Development', in *Minnesota Symposium on Child Development*, Vol. 11, ed. by W. A. Collins (Hillsdale: Lawrence Erlbaum, 1978), pp. 231–76; *Emotional Development*; *The Development of Behavioral States*.

36 *Emotional Development*; 'The Ontogenesis of Smiling and Laughter'.

37 C. E. Izard, 'Basic Emotions, Relations Among Emotions, and Emotion-Cognition Relations', *Psychological Review*, 99, 3 (1992), 561–65.

V. New Data on Neonatal Smiling in Pre-term and Full-term Infants

To further investigate smiling in early infancy, we conducted a series of studies partly supported by the *Ministero dell'Istruzione, dell'Università e della Ricerca* (PRIN 2004, 2004118439). The main purpose was to analyse the characteristics of the *zygomaticus major* muscle activity a few hours after birth. In particular, in association with some American (Miami and Utah University) and Italian (Padova and Calabria University) research units, we conducted four studies investigating the facial behaviour of independent groups of full-term [38] and pre-term neonates. [39] For the first time in literature these studies describe neonatal smiling using FACS and Baby FACS systems.

VI. Main Results and Broader Implications

Main results and theoretical implications arising from our research project may be summarized in four points.

VI.I. Duchenne Smiles in Pre-term and Full-term Neonates

First of all, by coding facial behaviour using Baby FACS, we found that Duchenne smiling is present in full-term neonates in more or less the same proportions as simple (non-Duchenne) smiling. [40] We also found that Duchenne smiling is frequent in pre-term neonates of 33/34 post-conceptional weeks [41] and even in

38 D. S. Messinger, Marco Dondi, C. G. Nelson-Goens, Alessia Beghi, Alan Fogel, and Francesca Simion, 'How Sleeping Neonates Smile', *Developmental Science*, 5, 1 (2002), 49–55; Marco Dondi, D. S. Messinger, Marta Colle, Alessia Tabasso, Francesca Simion, Beatrice Dalla Barba, and Alan Fogel, 'A New Perspective on Neonatal Smiling: Differences Between the Judgments of Expert Coders and Naïve Observers', *Infancy*, 12, 3 (2007), 235–55.

39 Marco Dondi, Angela Costabile, Manuela Rabissoni, Chiara Gianfranchi, Oreste Lombardi, and Carlo Corchia, 'Il sorriso del neonato prematuro: una ricerca esplorativa' (Smiles of Premature Neonates: An Exploratory Investigation), *Giornale Italiano di Psicologia*, 31 (2004), 377–99; Marco Dondi, Angela Costabile, Tiziana Vacca, Laura Franchin, Sergio Agnoli, Oreste Lombardi, and Carlo Corchia, 'Alle origini delle emozioni: uno studio pilota sul sorriso e sull'espressione facciale di distress nel neonato pretermine di età gestazionale molto bassa' (At the Origins of Emotions: A Pilot Study on the Facial Expression of Distress and Smiling in Very Low Gestational Age Preterm Newborns), *Psicologia Clinica dello Sviluppo*, 12, 1 (2008), 89–109.

40 'How Sleeping Neonates Smile'; 'A New Perspective on Neonatal Smiling'.

41 'Il sorriso del neonato prematuro'.

very immature infants of less than 30 weeks.[42] Surprisingly, the results revealed that more than one-half of pre-term neonates at 30 post-conceptional weeks exhibited Duchenne smiles bilaterally and at the highest muscular intensity level. This was unexpected in such a very low post-conceptional age. Previously some authors have reported anecdotal observations of the Duchenne smile in premature and in full-term newborns, but it was generally maintained that this particular smile was extremely rare at birth.[43]

Documenting the presence of Duchenne smiles in the neonatal expressive repertoire also suggests a different developmental trajectory of smiling than that described in literature. These findings actually contradict the widely shared notion that the emergence of *social smiling* coincides with the emergence of Duchenne smiling (see note 36 above), demonstrating that the morphology of the neonatal smile, even in very immature newborns, reflects an unexpected complexity.

The presence of Duchenne smiling raises past and present questions regarding the emotional meaning of this configuration at birth, which has important implications for the current theories on emotion and emotional development. The discovery of a mature, frequent, bilateral Duchenne smile is provocative, because it may suggest the precocious experience of enjoyment in neonates. If, however, Duchenne smiles are not joyful expressions during the neonatal period, we need to understand how they acquire positive emotional significance later in development. A central task for any systematic developmental analysis of human emotional expressions is describing and explaining how and when particular facial actions begin to refer to specific internal motivational states. Unfortunately, the current theories on the development of emotions have not yet provided a convincing account of this fundamental process.

VI.II. Smiling in Waking Neonates

The fact that the newborn infant frequently smiles also in an awake state constitutes another relevant result we obtained[44] with respect to what has been claimed in literature (see note 33 above), namely that neonates are capable of

42 'Alle origini delle emozioni'.

43 'Neonatal Smiling in REM States, IV: Premature Study'; 'Neonatal Smiling and Rapid Eye Movement States'; 'Neonatal Smiling, Frowning, and Rapid Eye Movement States. II Sleep-Cycle Study'; Harriet Oster, 'Facial Expression and Affect Development', in *The Development of Affect*, ed. by Michael Lewis and L. A. Rosenblum (New York: Plenum Press, 1978), pp. 43–74; *The Development of Behavioral States.*

44 'A New Perspective on Neonatal Smiling'.

smiling only during sleep. Combining micro-analytic anatomical coding of *zygomaticus major* activity with a smile recognition study, we found that smiling was not exclusively related to active sleep (irregular, paradoxical or REM sleep) and drowsiness, demonstrating that it was also relatively frequent during alertness (alert inactivity or quiet waking) and quiet sleep (regular sleep or non-REM sleep). Specifically, both lip corner raising (with or without cheek raising) coded by expert coders and recognized as smiling (Duchenne and non-Duchenne) by naïve observers occurred in all behavioural states examined. [45]

On the basis of these results, it appears necessary to abandon the idea, which has been assumed for a long time, that *zygomaticus major* activity is a prototypical behaviour related to REM state. This facial activity seems in effect to be related to subtle changes in the dynamics of behavioural states, but it is neither clearly nor specifically associated with active (REM) sleep, as previously maintained.

Interestingly enough, having demonstrated that newborn smiles during the awake state raise new questions regarding the functional relevance of early smiling, it could be hypothesized, for example, that very early smiling in waking states may represent a primitive or immature behaviour from which *social smiling* will gradually develop. This suggests continuity instead of discontinuity between neonatal smiling and early social smiling. The occasional smile that occurs during newborn waking states might lead to parents responding with behaviour such as smiles and vocalizations. Positive caregiver responses to these smiles might lead to reinforcing early social smiling, amplifying positive interactions with the infant. This mutual process of affective synchronization, which occurs in very early development along with a rapid growth of the nervous system, might lay the epigenetic foundations for the emergence of *social smiling*.

VI.III. Unexpected Characteristics of Neonatal Smiling

The use of micro-analytic methodologies such as FACS and Baby FACS, combined with the analysis of the smiling behaviour that is recognized by naïve observers, revealed another interesting result. [46] The same videotaped material was presented to expert facial coders (FACS coders) in a first study, and to naïve

45 Behavioural states were defined in terms of the conjunction of behavioural conditions, incorporating salient features of several well-known coding systems: T. B. Brazelton, *Clinics in Developmental Medicine, No. 88: Neonatal Behavioral Assessment Scale (NBAS)*, 2nd edn (Philadelphia: Lippincott, 1983); H. F. R. Prechtl and David Beintema, *The Neurological Examination of the Full Term Newborn* (London: SIMP Heinemann, 1964); *The Development of Behavioral States*.

46 'A New Perspective on Neonatal Smiling'.

observers (undergraduate university students) through a computerized smile recognition task in a second study. Contrasts between the data obtained in both studies demonstrated the presence of a brief *zygomaticus major* activity (with or without cheek raising) distinguished by expert coders but not recognized as smiling by naïve observers. In particular, expertly coded lip corner raising with or without cheek raising was typically briefer than the *zygomaticus major* activity recognized as smiling by naïve observers. Unexpectedly, in the second study the rate of neonatal smile occurrences was half the one recorded in the first study. This means that newborns displayed a certain amount of zygomatic activity that was not easily recognized as smiling. These muscular contractions were frequent and very brief.

On the basis of these data, it is clear that only a small portion of the zygomatic activity produced by the neonates is recognizable as smiling. From a developmental perspective this phenomenon is worthy of consideration, because the presence of a brief and low-level muscular activity in the *zygomaticus major* would be specific to the neonatal period. No data regarding this phenomenon are present to the best of our knowledge in developmental literature. Incidentally, this early fleeting activity does not appear dissimilar to the phenomenon of micro-expressions,[47] which are brief (sometimes visible only in one movie frame), involuntary and incomplete facial emotional expressions that occur on adults faces. However, contrary to what has been reported in adult studies, brief neonatal *zygomaticus major* activity appears to be very frequent.

A further characteristic recorded in both full-term and pre-term neonates was the presence of a recurrent unilateral contraction of the *zygomaticus major*, both in isolation and in synergy with the unilateral action of the ipsilateral orbicularis oculi. They represented, unexpectedly, more or less half of the entire amount of the *zygomaticus major* activity we recorded.

VI.IV. Neonatal Duchenne Smiling and Active (REM) Sleep

The smiling recognition by naïve observers and the coding performed by expert coders in the study described above revealed yet another important result. In both expert and naïve studies data demonstrated a slight but unambiguous relationship between active (REM) sleep and Duchenne smiling. In general, both lip corner raising along with cheek raising coded by expert coders and Duch-

47 Paul Ekman, *Emotions Revealed: Recognizing Faces and Feelings to Improve Communication and Emotional Life* (New York: Times Books, A Division of Henry Holt and Company, 2003); Paul Ekman and Maureen O'Sullivan, 'Who Can Catch a Liar?', *American Psychologist*, 46, 9 (1991), 913–20.

enne smiles recognized by naïve observers had a longer mean duration during active (REM) sleep than during the other behavioural states. In addition, it was only during active (REM) sleep that lip corner raising along with cheek raising and recognized Duchenne smiling had a longer mean duration respectively than lip corner raising alone and than recognized non-Duchenne smiling.

These findings suggest a relationship between active (REM) sleep and a relatively longer lasting Duchenne smile configuration. This association is challenging, because it is well known that REM sleep is characterized by intense neural activity, ocular saccades, muscular atonia, and dreaming along with limbic activation, which is typically involved in emotional responses. [48] Moreover, the role of active sleep (irregular, paradoxical or REM sleep) at birth is believed to be related to both structural and functional *central nervous system* (CNS) maturation, maintaining and establishing brain connections. [49]

For these reasons, neonatal Duchenne smiling exhibited during active (REM) sleep, more than those exhibited in other behavioural states as well as more than non-Duchenne smiling, may be influenced by limbic system discharges. As a working hypothesis, these longer lasting Duchenne smiling episodes may be conceptualized as expressions of the emerging coordination between a facial motor synergy (that is the association between the *zygomaticus major* and the *orbicularis oculi* activity) and the neural structures which are maintained to support an alert and social expression of smiling later in development, at about 2 months of age.

VII. Conclusions

The micro-analysis of early smiles in both pre-term and full-term newborn infants revealed unpredicted characteristics of this facial expression at birth, which has allowed for progress within the current state of scientific understanding. The availability of anatomically based facial coding systems (such as FACS and Baby FACS), as well as the introduction of improved technology (such as digital video-recordings) helped us to depict the emerging pattern of smiling in a rather different way compared to that reported during the Sixties.

48 B. E. Jones, 'Paradoxical Sleep and its Chemical/Structural Substrates in the Brain', *Neuroscience*, 40 (1991), 637–56; Pierre Maquet, J. M. Péters, Joel Aerts, Guy Delfiore, Christian Degueldre, André Luxen, and others, 'Functional Neuroanatomy of Human Rapid-Eye-Movement Sleep and Dreaming', *Nature*, 383, 6596 (1996), 163–66.

49 G. A. Marks, J. P. Shaffery, Arie Oksenberg, S. G. Speciale, and H. P. Roffwarg, 'A Functional Role for REM Sleep in Brain Maturation', *Behavioural Brain Research*, 69 (1995), 1–11; Majid Mirmiran, 'The Functions of Fetal/Neonatal Rapid Eye Movement Sleep', *Behavioural Brain Research*, 69 (1995), 13–22.

The frequent motor activity (bilateral and unilateral) of the *zygomaticus major* muscles (AU 12), with or without cheek-raising (AU 6) and across different behavioural states, sum up the main characteristics at birth of the facial pattern that specifies smiling. Sometimes this early motor activity is very brief and sometimes it is sufficiently enduring to be recognised as a meaningful facial expression. Unfortunately, current data do not permit us to establish whether neonatal smiling is a true expression of emotion or just a motor pattern still in search of a locus of control (neural, cognitive, social, emotional). Having this basic question as a reference, it will be a subject for future research to investigate the relationship between early episodic neonatal smiling in waking infants and *social smiling*, as well as the pattern of CNS activation which accompanies spontaneous smiling in different behavioural states.

Improvements in technology are already opening up new opportunities to investigate the very early development of facial expression of emotion. Recent three- and four-dimensional ultrasound scanning techniques in obstetric practice [50] are providing new prospects into the study of the facial behaviour emergence in prenatal life; entirely computerised systems for facial coding, such as the Automated Facial Image Analysis (AFIA), are now available. [51]

Nonetheless, main advancements will derive from the cross-fertilization between the numerous disciplines involved in emotion research. Neuro-scientists, for example, are building or improving neuro-imaging techniques suitable for neonates and infants. New methods of imaging the oxygenation, hemodynamics, and metabolism of the newborn infant brain are being developed. [52] The recent technological revolution in functional neuro-imaging has also disclosed new insights into regional brain activity during normal and pathological emotions and has advanced understanding of several neuropsychiatric diseases.

---

50 Luis Gonçalves, Wesley Lee, Jimmy Espinoza, and Roberto Romero, 'Three- and 4-Dimensional Ultrasound in Obstetric Practice: Does it Help?', *Journal of Ultrasound in Medicine*, 24 (2005), 1599–1624; Olivier Walusinski, Asim Kurjak, Wiku Andonotopo, and Guillermo Azumendi, 'Fetal Yawning Assessed by 3D and 4D Sonography', *The Ultrasound Review of Obstetrics & Gynecology*, 5, 3 (2005), 210–17.

51 J. F. Cohn, A. J. Zlochower, James Lien, and Takeo Kanade, 'Automated Face Analysis by Feature Point Tracking Has High Concurrent Validity with Manual FACS Coding', *Psychophysiology*, 36, 1 (1999), 35–43; K. L. Schmidt, Zara Ambadar, J. F. Cohn, and L. I. Reed, 'Movement Differences Between Deliberate and Spontaneous Facial Expressions: *Zigomaticus Major* Action in Smiling', *Journal of Nonverbal Behavior*, 30, 1 (2006), 37–52; Y.-l. Tian, Takeo Kanade, and J. F. Cohn, 'Recognizing Action Units for Facial Expression Analysis', *IEEE Transactions on Pattern Analysis and Machine Intelligence*, 23 (2001), 97–115.

52 Marco Bartocci, Jan Winberg, Gesa Papendieck, Teresa Mustica, Giovanni Serra, and Hugo Lagercrantz, 'Cerebral Hemodynamic Response to Unpleasant Odors in the Preterm Newborn Measured by Near-Infrared Spectroscopy', *Pediatric Research*, 50, 3 (2001), 324–30; J. C. Hebden, 'Advances in Optical Imaging of the Newborn Infant Brain', *Psychophysiology*, 40 (2003), 501–10.

The pattern of behavioural development we observe during early infancy may only provide an approximation of the neurobiological events that underlie changes in the emotional system. Neurosciences are now offering new insight and new information about the development as well as the functioning of the CNS. For these reasons we expect that our understanding of neural structures and circuits involved in emotional activation will achieve dramatic progress in the next few decades.

Laura Bafile

The Debate on the Origins of the Language Faculty in the Perspective of the Biolinguistic Programme

I. Introduction

Language as a property that distinguishes humans from animals and the way it emerged in our species have always aroused people's curiosity and often been a controversial subject among philosophers and linguists. In the last decades the debate about the origin of language has attracted renewed interest, also fuelled by knowledge from other fields like biology, genetics, ethology, and neurosciences.

The history of thought concerning language and its origin from the eighteenth century to the present day will be surveyed in relation to the major topics of current debate. The essential features of the biolinguistic programme and their implications on theories about the evolution of language will be illustrated and some controversial issues, especially related to the distinction between adaptationist and exaptationist approaches, will be addressed.

II. A Look at the History of the Debate

In 1866 the Société de Linguistique de Paris prohibited discussions about the origin and evolution of language: 'La société n'admet pas aucune communication concernant, soit l'origine du langage, soit la création d'une langue universelle'. [1] The Parisian ban is often mentioned as a reaction to the approaches to the origin of language dating to the early nineteenth and previous centuries, and to the lack of empirical evidence in related works.

However, both the intervention of the Société de Linguistique de Paris and the speculations about the origin of language formulated in the eighteenth and at the beginning of the nineteenth century deserve closer consideration. As Giorgio

1 Societé de Linguistique de Paris, 'Statuts approuvés par décision ministérielle du 8 mars 1866', *Bulletin de la Société de Linguistique de Paris*, I (1871), iii-v.

Graffi shows, on the one hand, the ban on investigation about origins is rooted in the paradigm which dominated linguistics in the nineteenth century; on the other hand, attempts in previous centuries to explain the origin and evolution of language actually anticipated some of the crucial issues of current debate. [2]

In the eighteenth century the question about origins shifted from the search for the original tongue, mother of all existing languages, to language meant as a faculty of the human mind. The conception of language as 'analytical tool of thought' advanced in 1660 in the Port-Royal *Grammaire* was developed in the following century by Condillac, one of the most important philosophers participating in the debate about the origin of language.

During the nineteenth century, when the historical-comparative paradigm was predominating, the main interest of linguistic research shifted again from *langage*, 'language' in the sense of cognitive capacity, to *langue*, 'language' in the sense of a system shared by a community. As a consequence, as Graffi highlights, the problem of the origin of language was regarded as a pseudo-problem. The prohibition of the Société de Linguistique de Paris actually reflects, and ratifies, this perspective.

In the structuralist period, during the first half of the twentieth century, research mainly focused on the constitutive elements of linguistic systems, and a philosophical question like the origin of language was virtually ignored.

Only since the last decades of the twentieth century language as mental faculty, the I(nternal)-language in Chomsky's terms, has become again a crucial object of linguistic investigation, and the debate about origins has resurfaced. Pinker's and Bloom's *Natural Language and Natural Selection*, [3] published in 1990, is the first study which adopts an evolutionary approach to language based on Chomski's theory of generative grammar.

As noted above, some issues of contemporary debate were anticipated in the seventeenth and eighteenth centuries. Speculations as to whether prelinguistic forms of expression like gestures and cries could be regarded as precursors for language do not sound very different from some of the hypotheses assumed by contemporary scholars. [4] In the same way, Condillac's idea of the *language d'action* as the first form of language, consisting of single words and two-word units, continues, as Graffi points out, in the contemporary hypothesis of *pro-*

2 Giorgio Graffi, 'The Problem of the Origin of Language in Western Philosophy and Linguistics', *Lingue e linguaggio*, 6, 1 (2005), 5–26.

3 Steven Pinker and Paul Bloom, 'Natural Language and Natural Selection', *Behavioral and Brain Sciences*, 13, 4 (1990), 707–26.

4 See for example Chris Knight, 'Introduction: Grounding Language Function in Social Cognition', in *Approaches to the Evolution of Language: Social and Cognitive Bases*, ed. by James R. Hurford, Michael Studdert-Kennedy, and Chris Knight (Cambridge: Cambridge University Press, 1988), pp. 9–16.

tolanguage. [5] More generally, generative approaches to the problem of the origins recall Condillac's and Steinthal's assumption that the explanation of the origin of language, on the one hand, and the analysis of grammar and acquisition of natural languages, on the other, need to be convergent.

Pinker and Bloom's 1990 paper resumed scientific interest in the origin of language and raised a lively debate, which has continued ever since, confronting generative linguistics and anti-Chomskian positions.

Linguists who support Chomsky's theories consider language as an innate, species-specific, biologically-determined faculty. This conception is based on empirical considerations of various kinds: [6]

a. All human societies have language.
b. All natural languages are based on the same kinds of principles, which are not notably connected with technological progress: the grammars of languages used in modern industrial society are not more complex than the grammars of prehistoric societies. For example, modern English is not a language more advanced than Old English.
c. Individuals have native competence of a language, regardless of intelligence, social status and education.
d. During the processes of transformation of pidgins into creole languages, children prove capable of inventing languages similar to languages they have never heard, respecting grammatical principles of which there is no evidence in their linguistic environment.

Nevertheless, discussion is also lively among generative linguists, who support two main arguments. On the one hand, Bickerton, Pinker, and Jackendoff, among others, [7] aim at an adaptationist explanation, trying to apply to language the Darwinian theory of evolution, 'in which complex biological systems arise by a gradual accumulation over generations of random genetic mutations that enhance reproductive success'. [8] On the other hand, Chomsky and others reject the hypothesis whereby the complex of interacting capacities that activate human language might be the result of a single evolutionary design. In this view, sometimes defined exaptationist, the core linguistic capacity, that consists in merging symbols in order to form a boundless system of structured expressions,

5 See Dereck Bickerton, *Language and Species* (Chicago: University of Chicago Press, 1990).
6 This summary is taken from 'Natural language', pp. 707–08.
7 See for instance *Language and Species*; Steven Pinker, *The Language Instinct* (New York: Morrow, 1994); Ray Jackendoff, *Foundations of Language: Brain, Meaning, Grammar, Evolution* (Oxford: Oxford University Press, 2002).
8 *The Language Instinct*, p. 333.

may have emerged and then affirmed in the course of the evolution for reasons other than merely linguistic ones. [9]

Anti-Chomskian theories are based on arguments of different kinds. Some of them support the continuist hypotheses that identifies communication systems of monkeys and apes as the precursors of human language. Others consider language as a manifestation of general cognitive capacities and attribute to socio-communicative and cultural factors a crucial role in the development from more rudimentary forms of communication to modern languages. [10] A particular perspective sees the emergence of grammar as a product of cultural transmission of language across generations, a process that is considered the cause of increasing complexity in linguistic systems. Evidence supporting this hypothesis is said to be grammaticalization, a drift observed in the history of languages, whereby words tend to become grammatical elements, such as *legere habeo* in late Latin and *leggerò*, (<*leggere ho*),'I will read', in Italian. [11]

III. The Biolinguistic Approach to Language

The specific question of the origin of the language faculty receives only secondary attention in Chomsky's investigation into the architecture of language. Until Pinker and Bloom's published their article, this subject was scarcely treated by generative linguists. Nevertheless, Chomsky's conception of language and the topics he has investigated since the beginning of generative linguistics have raised new interest in the problem of origin and stimulated new approaches.

Chomsky, also encouraged by the work of neuropsychologist Eric Lenneberg, author of *Biological Foundations of Language* (1967), [12] regarded the faculty of language as an organ of the body, a biological system. This gave rise to a new perspective, later called 'biolinguistics'. By interacting since the 1950s 'with a more general shift of perspective in the study of mental faculties, commonly

9 See for instance Noam Chomsky, *Language and Problems of Knowledge: the Managua Lectures* (Cambridge, MA: MIT Press, 1988); Massimo Piattelli-Palmarini, 'Evolution, Selection and Cognition: from 'Learning' to Parameter Setting in Biology and in the Study of Language', *Cognition*, 31, 1 (1989), 1–44.

10 See for instance Chris Knight and Maria Ujhelyi, 'Long Call Structure in Apes as a Possible Precursor for Language', in *Approaches to the Evolution of Language*, pp. 171–77.

11 See Michael Tomasello, 'On the Different Origins of Symbols and Grammar', in *Language Evolution. The States of the Art*, ed. by Morten H. Christiansen and Simon Kirby (Oxford: Oxford University Press, 2003), pp. 94–110; Bernd Heine and Tania Kuteva, 'On the Evolution of Grammatical Forms', in *The Transition to Language*, ed. by Alison Wray (Oxford: Oxford University Press, 2002), pp. 376–97.

12 Eric H. Lenneberg, *Biological Foundations of Language* (New York: Wiley, 1967).

called 'the cognitive revolution", [13] the biolinguistic programme has determined a deep change in the approach to the study of language and mind. While in the 1950s such study was commonly considered part of behavioural sciences, in the new perspective it has become part of natural sciences, with which it shares the organization of research on the basis of standards and procedures of deductive nature. [14]

According to the fundamental principles of generative grammar, the faculty of language is a biologically isolated property, since it is exclusively found in the human species. Other organisms may have their systems of communication which, however, have radically different properties. In particular, an essential property of human language, namely discrete infinity, is rare in the organic world. This is the capacity of merging words into phrases and phrases into sentences, or to produce expressions of potentially unlimited length by recursively inserting a sentence into another.

The specific object of the scientific study of language is the internal language (I-language), that is a system of knowledge represented in the mind/brain of a particular individual. This position takes into account Chomsky's critique of behaviourist speculations about the nature of language and the way it is acquired by children: no account of speakers' and listeners' behaviour or of language acquisition is possible unless in terms of an in-built structure that determines the growth of grammar through exposition to linguistic stimuli. [15]

A distinctive property of the cognitive system of language is its specificity, i. e. the fact that it is grounded on specifically linguistic principles. Many syntactic phenomena in natural languages do not find any explanation in the principles of logic or general cognition. [16] To give an example, to derive the interrogative sentence in (1b) from the declarative in (1a), we could at first glance put forward as an hypothesis the rule that the verb *is* must be moved to the beginning of the sentence. However, the impossibility of deriving (1d) from (1c) shows that the actual operation of movement applies to the main verb, which is the verb that agrees with the subject of the main sentence:

13 Noam Chomsky, 'Universals of Human Nature', *Psychotherapy and Psychosomatics*, 74 (2005), 263–68 (p. 264).

14 See Noam Chomsky, *Rules and Representations* (New York: Columbia University Press, 1980); Id., *Knowledge of Language: Its Nature, Origin and Use* (New York: Praeger, 1989); Id., 'Three Factors in Language Design', *Linguistic Inquiry*, 36, 1 (2005), 1–22.

15 Noam Chomsky, 'A Review of B. F. Skinner's Verbal Behaviour', *Language*, 35, 1 (1959), 26–58.

16 See 'Evolution, Selection and Cognition'; Massimo Piattelli-Palmarini, 'Ever Since Language and Learning: Afterthoughts on the Chomsky-Piaget Debate', *Cognition*, 50 (1994), 315–46.

(1)
a. The man is here
b. Is the man here?
c. The man who is tall is here
d.* Is the man who tall is here?

The rule that produces the correct sentence *Is the man who is tall here* refers to entities like 'subject', 'main sentence', that have strictly linguistic and structural relevance; it is a so-called 'structure-dependent' rule.

Evidence of specific linguistic principles can also be found by comparing languages. Examples in (2) and (3) show different properties of interrogatives in Italian and English. In Italian the interrogative element *chi* 'who', moved to the beginning of the sentence (see 2c), may substitute both the object (see 2a) and the subject (see 2b), while in English only the object may be substituted by *who(m)* and moved to the beginning; hence the incorrectness of (3b'):

(2)

a.	Pensi	che	John	sposerà	qualcuno
	You think	that	John	will marry	someone
b.	Pensi	che	qualcuno	sposerà	John
	You think	that	someone	will marry	John
c.	Chi	pensi	che	sposerà	John?
	Who	you think	that	will marry	John?

(3)
a. You think that John will marry someone
a'. Who(m) do you think that John will marry?
b. You think that someone will marry John
b'.* Who do you think that will marry John?

The difference between Italian and English shown in (2) and (3) as well as other apparently unrelated syntactic phenomena of both languages depend on the fact that the subject must be obligatorily expressed in English and not in Italian. More generally, what data like those in (1 – 3) show is that many properties of the grammars of natural languages can only be described in the specific terms of linguistic structure. For the same reasons, in order to be learnt and realised in linguistic expressions, those properties imply the existence of a set of specific linguistic principles in human mind.

The growth of language in the individuals, however, obviously implies two other factors, besides the genetic endowment of specific linguistic principles:

i) experience; ii) principles not specific to the faculty of language. Chomsky describes the interaction among the three factors as follows:

> [...] the early steps of compiling linguistic experience might be accounted for in terms of general principles of data analysis [*i.e. principles not specific to the language faculty*] applied to representations preanalyzed in terms of principles specific to the language faculty [...]. [17]

The importance attributed to each of the three factors has been changing in the history of generative grammar. In earlier approaches, the genetic component was considered to be highly articulated and richly structured, so as to explain learnability of grammars. In the 1980s the Principles-and-Parameters framework determined a radically different approach to the nature of biological endowment and acquisition of language: [18] the genetic component of the language faculty, the Universal Grammar, consists of a limited set of universal abstract principles, that determine universal properties of languages, and in 'menus' of alternative options, called parameters, that determine the variation among languages. To give an example, the obligatory expression of the subject in a sentence, which distinguishes the group of languages that behave like English from the group of languages that behave like Italian, depends on a parameter, on which a cluster of other syntactic phenomena also depends. [19] Within this framework, many properties of the grammar of individual languages depend on parametric options, and acquisition is a matter of parameter setting. To continue with the example concerning the subject, setting the proper value of the Null Subject parameter ('Subject may be omitted: YES/NO') is what a child needs in order to achieve full competence of the complex of properties, some of which exemplified in (1 – 3) above, concerning his/her mother tongue.

Given the Principles-and-Parameters framework, the genetic linguistic component of language faculty may be reduced to a form much simpler than deemed in earlier approaches: a few, very abstract principles. In fact, more specific features of different languages must be acquired through parameter setting.

In the last decade, an even more radical simplification has characterised Chomsky's and other generative linguists' view of the core grammar, and therefore of the genetic linguistic endowment. In two studies concerning the nature of language faculty and its degree of specificity in human species,

17 'Three Factors', p. 7.

18 See Noam Chomsky, *Lectures on Government and Binding* (Dordrecht: Foris, 1981); Mark C. Baker, *The Atoms of Language* (New York: Basic Books, 2003).

19 See Luigi Rizzi, *Issues in Italian Syntax* (Dordrecht: Foris, 1982); *The Atoms of Language*; *Lectures on Government and Binding.*

Chomsky, Fitch, and Hauser have collected evidence of different kinds for a distinction between the faculty of language in narrow sense (FLN) and the faculty of language in broad sense (FLB). [20] They have proposed that the specifically human capacity, FLN, may only consist in discrete infinity, i.e. the capacity 'of recombining meaningful units into an unlimited variety of larger structures, each differing systematically in meaning'. [21] The hypothesis is that discrete infinity is a uniquely human property, which, however, has not evolved as a unique adaptation for communication, since it also characterises natural numbers and possibly other components of human cognition.

With respect to the question of origins, the simplification of the genetic endowment specific for language elicits an easier account of the evolution of language. [22] What needs to be explained in evolutionary terms is the single, even though crucial, trait of FLN, and not FLB, i.e. the complex of interacting capacities that allow for the use and the acquisition of language.

Also worthy of attention in Chomsky's view is the difference between language and communication. [23] Although languages are a typical instrument of interaction among humans, they are much more than a mere communication system. The equivalence between communication and language is based on the wrong assumption that both are manifestations of a single code. A crucial argument in this regard is that some structural properties of languages are neither optimal nor fully efficient for the purposes of communication. [24]

One of such properties is the movement required, for example, by passivization or interrogative constructions. To interpret a sentence like (3a'), here repeated in (4), it is necessary to locate the object of *marry*, expressed by *who*, in its basic position after the verb:

(4)
Who(m) do you think that John will marry [who(m)]?

Syntactic movement undoubtedly makes the process of speech perception more complex, and therefore hinders communication.

Similar considerations hold for discrete infinity, since the potentiality of inserting sentences into other sentences clearly clashes with communicative

20 Marc D. Hauser, Noam Chomsky, and W. Tecumseh Fitch, 'The Faculty of Language: What Is It, Who Has It, and How Did It Evolve?', *Science*, 298 (2002), 1569–79; W. Tecumseh Fitch, Marc D. Hauser, and Noam Chomsky, 'The Evolution of the Language Faculty: Clarifications and Implications', *Cognition*, 97 (2005), 179–210.

21 'The Faculty of Language', p. 1576.

22 'Three Factors'.

23 See for instance *Knowledge of Language*.

24 'Ever Since Language and Learning'.

efficiency. As a sentence like (5) shows, syntactic recursivity generates expressions that, despite their grammatical correctness, are far from optimal for communication; for this reason, this capacity is exploited in a very limited way in the actual use of language:

(5)
I remember that John told me that Paul believes that Mary wants that Chris marries Paul

Moreover, ambiguity is a common property of linguistic expressions, that obviously clashes with communicative efficiency. Consider a sentence like (2c), repeated in (6a), and the structures corresponding to each of the possible meanings (6b, 6c):

(6)

a. Chi	pensi	che	sposerà	John?
Who	you think	that	will marry	John?

b. Chi pensi che [chi] sposerà John?
'Who do you think will marry John?'
c. Chi pensi che sposerà [chi] John?
'Whom do you think John will marry?'

Research in the field of pragmatics, especially by Sperber and Wilson, has shown that linguistic communication needs a linguistic code, but does not simply consist in coding and decoding operations, whereas it crucially involves operations of inference. [25]

The distinction between language and communication obviously affects speculations about the origin of language. If, in its narrow sense, the faculty of language is not designed for communication, it cannot have evolved for communicative functions. In recent works Chomsky has observed that, since language is an instrument of thought, externalization, i.e. translation of conceptual-linguistic structures into sounds, must be a secondary process. His hypothesis is that externalization, and therefore interaction and communication, might have emerged in evolution after the essential principle of language faculty, the core capacity of combinatorial productivity, had emerged in an individual as a consequence of some rewiring of the brain. [26] The individual so endowed would have known a dramatic increase of cognitive capacities, that

25 Dan Sperber and Deirdre Wilson, *Relevance: Communication and Cognition* (Blackwell: Oxford University Press, 1986; 2nd edition 1995).
26 'Three Factors'.

were transmitted to offspring and determined what the paleoanthropologist Ian Tattersall calls 'creative explosion', i. e. the sudden transition from a non-symbolic to a symbolic culture, corresponding to the invention of language. [27]

IV. The Uniqueness of Language and Other Controversial Issues

As we have previously noticed, one of the fundamental arguments of the biolinguistic programme is that capacity for language (which in generative tradition has been variously defined as Language Acquisition Device, Universal Grammar, 'language organ', or 'language instinct') is built into the human brain. A generative theory of language evolution must therefore account for the presence of the genetic trait underlying linguistic capacity. For this reason, defining the nature of this capacity and understanding to what extent it is specific for language is a task of linguistic theory that has crucial implications for any hypothesis about language evolution. Fundamental questions are:

Which aspects of the language faculty are specific for humans and which ones are shared with other animals?

Which aspects of the language faculty are specially required by language and which ones are part of more general capacities?

The answers to these questions will also clarify to what extent the faculty of language implies abilities that are new in evolutionary terms, i. e. that do not have precursors. Other things being equal, a more parsimonious theory of language that presupposes as few specific and unprecedented elements as possible is to be preferred on evolutionary grounds.

Chomsky, Fitch, and Hauser have proposed a distinction between the faculty of language in broad sense and the faculty of language in narrow sense. FLB is a 'biological capacity of humans that allows us (and not, for example, chimpanzees) to readily master any human language without explicit instruction', while 'FLN is the abstract linguistic computational system alone, independent of the other [*i.e. sensory-motor and conceptual-intentional*] systems with which it interacts and interfaces'. [28]

The implication of this minimalist conception of linguistic capacity in narrow sense is that language evolution coincides with the emergence of a single trait, the capacity for recursion. As noted above, externalization of this capacity is considered by Chomsky as a secondary trait, that, after recursion, may have emerged due to the advantages coming from interaction and communication.

27 Ian Tattersall, *Becoming Human: Evolution and Human Uniqueness* (New York: Harcourt Brace, 1998).

28 'The Faculty of Language', p. 1571.

A fundamental role in answering the questions posed above is played by comparative research, which investigates the existence of abilities essential for human language in animals and compares linguistic capacities with other perceptive and cognitive skills of human mind.

Animal capacity for (some aspects of) language has been investigated since the 1950s. The first approach, defined as 'training studies', has researched the possibility of teaching primates spoken English or some other symbolic system; the second approach involves the observation of communication among animals in natural environment.

The goal of training studies clearly sounded exciting from the evolutionary perspective: if the uniqueness of human language could be challenged, an hypothesis about language could be formulated that was consistent with Darwin's theory of gradual evolution by natural selection. [29]

Attempts to teach apes to speak English were complete failure: after six years' training, the vocabulary used by Vicky, a chimpanzee *(Pan troglodytes)* instructed in spoken English, consisted of no more than 4 words (*mama, papa, cup, up*). [30] The reason for this failure was found in the anatomy of the vocal tract of primates, that makes them unable to utter the variety of linguistic sounds. Therefore, when in 1966 Allen and Beatrice Gardner started their project with Washoe, a one-year-old chimpanzee, they decided to teach her American Sign Language. After three years Washoe had learnt 85 signs, and after ten years she was reported to know 160 signs. [31]

In the 1980s Savage-Rumbaugh started her research on the bonobo chimpanzee *(Pan Paniscus)* Kanzi, and her work has produced somewhat more encouraging results. [32] Kanzi is reported to be able to use hundreds of lexigrams, i. e. geometric symbols drawn on a board, to understand numerous words of spoken English, and to access conceptual notions like agent, object and location.

On the whole, however, data resulting from training studies show that, although these animals may be able to use symbols, they do it through processes that are fundamentally different from those underlying lexical and constituent structure. Even when they combine two or three symbols, they do it in a very

29 See Marc D. Hauser, David Barner, and Tim O'Donnell, 'Evolutionary Linguistics: A New Look at an Old Landscape', *Language, Learning and Development*, 3, 2 (2007), 100–32.

30 See Cathy Hayes, *The Ape in Our House* (New York: Harper & Row, 1951).

31 See Allen R. Gardner and Beatrice T. Gardner, 'Evidence for Sentence Constituents in Early Utterances of Child and Chimpanzee', *Journal of Experimental Psychology*, 104 (1975), 244–67.

32 E. Sue Savage-Rumbaugh, *Ape Language: From Conditioned Response to Symbol* (New York: Columbia University Press, 1986).

restricted way, never constructing anything similar to clauses or sentences, with respect to structure and expressive power. [33]

A different approach to the question of language in animals consists in the exploration of communication in natural environment. Calls of different kinds in monkeys, apes and other species turn out to be very effective communication systems, that however do not exhibit the basic properties of human language: they are not open-ended systems, they never refer to objects and events out of the 'here and now' (with the exception of bee's dance language), they do not appear to include abstract categories like verb or noun. [34]

In recent years a different approach has emerged concerning the uniqueness of human language, i. e. the comparative approach. In comparative perspective, neither animals' ability to use some sort of human language, nor animal natural communication systems are investigated. Instead, the question is which of the abilities implied by language in human beings may also be found in animals, for communicative or other purposes. The comparative perspective is based on a non-holistic view of the language faculty: language is not a monolithic capacity, and implies a cluster of different, though interacting, abilities.

The problem of uniqueness is obviously a crucial question concerning language evolution. If no evidence of a precursor mechanism is found in other animals, the abilities required by language in humans are likely to be new evolutionary traits, probably evolved for linguistic reasons. Interesting comparative data result from research in the field of speech perception and production.

In the 1960s Liberman and his colleagues at the Haskins laboratories began to explore in details the mechanisms of speech perception, and thus discovered the phenomenon of categorical perception. [35] This is a perceptive mechanism whereby differences among items that fall into different categories are enhanced, while differences among items that fall into the same category are minimized. When we perceive speech, we create categories. In a continuum from unvoiced to voiced consonants, for example from [pa] to [ba], human adults apply a categorical discrimination, which is excellent for the between-category exemplars, but poor for the within-category exemplars.

The first hypothesis was that categorical perception is a human specifically linguistic mechanism of perception. At that time, when biolinguistic research programmes were starting, the discovery of categorical perception seemed to reinforce the idea of an innate, species-specific language faculty. In following

33 See Edward T. Kako, 'Elements of Syntax in the Systems of Three Language-trained Animals', *Learning & Behavior*, 27, 1 (1999), 1 – 14.

34 'Evolutionary Linguistics'.

35 See A. M. Liberman and others, 'Perception of the Speech Code', *Psychological Review*, 74 (1967), 431 – 61.

years, however, comparative investigation has shown that categorical perception also concerns non-speech stimuli in humans and is an auditory process shared with other animals (like chinchillas, macaques and birds). This mechanism has not evolved for speech, and may have evolved much earlier than the emergence of language, before the divergence point with birds. [36]

One of the abilities required by speech production is vocal imitation, i. e. the ability to reproduce complex sounds vocally. This ability would appear to be exclusive to humans among primates, and therefore to be a candidate for a uniquely linguistic property, but comparative research has found robust evidence of vocal imitation in most songbirds and some marine mammals. [37]

Moreover, as Fitch, Hauser, and Chomsky observe, [38] human imitation is not specific for language; all humans can sing and easily reproduce a novel tune. Their conclusion is thus that vocal imitation is not included in the faculty of language in narrow sense, and has not evolved for language.

An anatomical issue regarding adaptation for speech production is the question of descended larynx. Experiments with chimpanzees failed because of the difference in the anatomy of vocal tract. In human species, larynx has a lower position than in primates, and therefore it enables the production of a richer variety of sounds. Given its absence in primates, and its function in language, descended larynx was believed to be a uniquely human trait, evolved for increasing speech versatility.

However, recent comparative studies have discovered descended larynges in other animals. The most plausible adaptive explanation for the larynx descent in animals is the 'size exaggeration' hypothesis, whereby the lowered position of larynx allows for impressive roars that can intimidate rivals and attract females. Size exaggeration may also be an adaptive explanation for larynx descent in our own species, particularly in the light of the fact that after an initial descent observable in infants, a second descent happens in puberty, but only in males: significantly, this second descent does not increase phonetic versatility, but increases the impressiveness of the adult male voice. [39]

Many results of comparative research until now appear to be consistent with Chomsky's hypothesis that, in the evolution of language faculty, the only uniquely human (but possibly not uniquely linguistic) trait is the capacity for recursion. However, this position is highly controversial also among scholars of generative grammar.

In 2005 Jackendoff and Pinker rejected Chomsky's, Fitch's and Hauser's

36 See Marc D. Hauser and Tecumseh W. Fitch, 'What Are the Uniquely Human Components of the Language Faculty?', in *Language Evolution*, pp. 158–81.

37 Ibid., ivi.

38 'The Evolution of the Language Faculty', p. 198.

39 'What Are the Uniquely Human Components'.

conception of language and language evolution in many respects. [40] First, FLN does not consist only of syntactic recursion. Secondly, language evolution may well have been a gradual process. Furthermore, interaction and communication must have had a determining role as adaptive factors in the evolution of language.

Jackendoff and Pinker argue that aspects of language other than recursion must be included in FLN. This claim is also due to a different interpretation of the 'uniqueness' of a trait. In absolute terms, a trait is unique (unique to humans, unique to language) if there is 'nothing remotely similar in the rest of the animal kingdom' and it is 'appearing out of the blue in the course of evolution'. [41] In the 'graded' interpretation of Jackendoff and Pinker, a trait can be considered unique also when it 'has been modified in the course of human evolution to such a degree that it is different in significant aspects from its evolutionary precursors'. [42]

In the 'graded' interpretation, any aspect adapted for language, even if deriving from an evolutionary precursor, is part of FLN. For example, while humans admittedly perform vocal imitation of noises and melodies, they do it with much inter-individual variance, while the imitation of linguistic sounds is a fine-grained ability, observable in all the individuals of the species, and can, therefore, be included in FLN.

In a 2008 manuscript Jackendoff claims that Chomsky's linguistic theory has a 'syntactocentric' architecture, where words are associations of phonological, syntactic, and semantic features, inserted into a syntactic structure when sentences are made up. In this view, syntax is the only combinatorial component. [43] Instead, Jackendoff has proposed a 'parallel' architecture, in which the different components of phonology, syntax, and semantics, based on independent principles of combinatoriality, interface with each other. [44]

One of the implications of parallel architecture is that, being the semantic/conceptual structure a product of a combinatorial capacity independent of syntax, our ancestors could have had highly structured thought before they had language. Moreover, parallel architecture allows for a direct correspondence

40 Steven Pinker and Ray Jackendoff, 'The Faculty of Language: What's Special about It?', *Cognition*, 95 (2005), 201–36; Ray Jackendoff and Steven Pinker, 'The Nature of the Language Faculty and Its Implications for Evolution of Language (Reply to Fitch, Hauser, and Chomsky)', *Cognition*, 97 (2005), 211–25.

41 'The Nature of the Language Faculty', p. 214.

42 Ibid., ivi.

43 Ray Jackendoff, 'Your Theory of Language Evolution Depends on your Theory of Language', *ms.*

44 See *Foundations of Language*; Peter W. Culicover and Ray Jackendoff, *Simpler Syntax* (Oxford: Oxford University Press, 2005); Peter W. Culicover and Ray Jackendoff, 'The Simpler Syntax Hypothesis', *Trends in Cognitive Science*, 10, 9 (2006), 414–18.

between phonology and semantics. This opens up the possibility of a 'hominid paleo- or proto-lexicon' consisting of pieces of thought associated with vocal or gestural expressions.

Jackendoff's hypothesis is that proto-lexicon could permit the expression of thought with no intervention of anything like modern syntax. In fact, linearization in these expressions could depend on rules based on semantic relations, such as 'Agent first, Focus last', rather than on syntactic requirements.

This syntax-free communication system roughly corresponds to Bickerton's *protolanguage*. He has proposed that language faculty evolved in two stages, and that the first was protolanguage. According to his hypothesis, protolanguage is still present in modern humans, surfacing in particular circumstances like language acquisition, syntactic deficit, or, in situations of linguistic contacts, in the form of a pidgin. [45] Pidgins are contact codes, that no speaker has as his/her mother tongue, and are characterised by an almost absent syntax. For this reasons, pidgins are sometimes considered linguistic 'fossils', in that they are believed to testify to a stage in the evolution of language. [46]

On the basis of parallel architecture and the protolanguage hypothesis, language evolution may be conceived as a gradual process, in which linguistic capacity increases in stages, each corresponding to an improvement in communicative efficiency. On the whole, this appears a rather different scenario from the hypothesis of Chomsky, Tattersall, and others, who consider the invention of language a prerequisite for the 'great leap forward' in the cognitive and cultural evolution of human species.

45 *Language and Species.*

46 See Rudolph Botha, 'Pidgin Languages as a Putative Window on Language Evolution', *Language and Communication*, 26 (2006), 1 – 14, for a discussion of the concept of pidgins as 'windows on language evolution'.

Keith Hoskin

Origins or Geneses? The Roles of Invention

I. Introduction: Why Geneses?

'Origins' and 'geneses' are both well-formed words; but the former is a more familiar plural form than the latter. That I prefer to begin from the less familiar 'geneses' here is because of a general question raised by Michel Foucault, concerning the appropriate way to describe or characterise human 'beginnings'.

Foucault named the first great series of his works, culminating in *Les Mots et les Choses* (1966) (in English *The Order of Things*, 1970) as 'archaeological': the second series he named as 'genealogical'– with the most recognised work probably being *Surveiller et Punir* (1975) (in English *Discipline and Punish*, 1977). In making this terminological transition he was, I suggest, doing two things.

First he was recognising that he had exhausted one methodological approach, a digging down to understand a given era and its discourses and practices by excavating its 'strata'; he found himself instead, in his intellectual journey, wanting to engage more directly with the how and why of human change; since given strata could be seen to be sedimented on older ones and then to give way to newer ones, there had, given that strata construction was a human endeavour, to be a human 'how' and 'why'. That did not mean that such hows and whys of change were necessarily rationally 'willed' or foreseen in advance, nor that changes would necessarily turn out as envisaged beforehand. Indeed one of Foucault's great historical contributions, in both his archaeological and genealogical works, was to demonstrate just how far radical shifts had to be involved – to the point where they could sometimes be named as 'discontinuities' – for strata to change at all.

Second he was, in seeking to understand such discontinuities, bringing into the foreground what is entailed in any form of human change and how it may best be characterised. 'Genesis' as a characterising term kept in view, more than origins, how there is firstly always some set of acts undertaken by the living prior to and absolutely necessary to the launching or initiating of what subsequently comes out of concealment as something different, and which 'becomes' separately, distinctly and beyond what was living before; genesis simultaneously allows for how what 'becomes' separately, distinctly and beyond, does so while remaining derived *from*, and so connected and related *to*, what 'became' previously. I choose terms here which will hopefully bear the weight of referring equally to biological and historical forms of genesis, particularly in relation to 'human' beings and their becoming.

If I seek, at the same time, to focus here more on the historical (or historical-biological) rather than on the purely biological forms of human genesis, that is because I would also wish to argue that in the world of human temporal action (including in action, thought), when there is some breakthrough or change that we may in retrospect be able to specify as a moment when new modes of human becoming materialise, we may well discover that there has been some form of 'invention'. In other words, such forms of genesis entail the creation or construction of something that was not before, but which either subsequently proves to have been the launchpad for whole series of variations on its new theme, or which becomes so integral to human becoming that it may widely be perceived as having been there for ever. Invention in this sense is transitive: it takes us as humans with it, so that we internalise it as what is now integral to our future becoming; we cannot go back to the world as it was before. But it is, simultaneously, never rationally 'willed' or foreseen in advance: for if it could have been, it would have been invented earlier. The classic example is perhaps writing: before the invention of writing, in around 3200 BCE, there are no visibly distinguished and distinct linguistic and numerical signs, so 'literate' thinking is simply an absence, as much work has long demonstrated. [1] We think in different ways, in a different world.

Now a simple distinction of writing-based 'literacy' from pure 'orality' can no longer be sustained, given the work of Schmandt-Besserat, [2] which shows that there are, from around 8000 BCE, systems of visible signs used for accounting in the first long-term successful agricultural and city-based economies and states, and that the first systems of writing developed in Mesopotamia, involving sets of separate linguistic and separate numerical signs, can be extrapolated semiotically and materially from these earlier systems. At the same time, the same principle obtains. Here was simply an earlier invention, in this case of a set of visible signs each of which was always *simultaneously* linguistic and numerical, and which, like all forms and systems of accounting down to today, therefore

1 Cf. Walter Ong, *Orality and Literacy* (London: Methuen, 1982); Jack Goody, *The Logic of Writing and the Organization of Society* (Cambridge: Cambridge University Press, 1986).
2 Denise Schmandt-Besserat, *Before Writing*, 2 vols (Austin: University of Texas Press, 1992).

always *simultaneously* 'named and counted' things. [3] In each case, the invention was not rationally willed or foreseen in advance, but once it had stolen out of concealment, it took us humans with it, as something internalised and integral to all who became versed in its use and extension, in a transitive way so that such humans could not go back to the world as it was before – nor indeed could those who remained unversed in the sign system as such but who came to live under the economic and political regimes that each made possible. So for instance, before the successful invention of sustained agriculture, all humans are hunter-gatherers. But once just a few humans have developed the settled home (in Greek the *oikos*) whose life is regulated by some system of produce allocation (in Greek *nomos*), all those living the new 'oikonomic' life, whether skilled in token use or not, live differently. Equally those beyond the world of the '*oikos*', such as the 'nomads' celebrated in the 'nomadology' elaborated by Deleuze and Guattari in *A Thousand Plateaus* (1988), only become definable as 'nomads' (or wanderers) once there is a category of people who no longer live on the move, but who redefine human living as something 'inherently' definable in terms of the 'settled oikonomic life'. At which point 'nomads' become definable as other and threatening, as drivers of what Deleuze and Guattari describe as 'The War Machine', [4] even if all that they do, through continuing to choose a way of life as 'other' to that ruled by token accounting, is exist outside the *oikos* and so to simply 'be' a potential 'threat' to oiko-nomic living, whether they choose, as supposed 'War Machine', to attack particular 'oiko-nomic' settlements or not.

In all the literate societies that follow down to the nineteenth century, the non-literate majority continue to be subject to the political and taxing powers of the state, as exercised through the particular forms of accounting and writing and numeration utilised at each particular time and place. And always, too, there are analogues to the nomads, who live some form of counter-life to that named and counted through accounting, writing and numeration. From ancient anarchists to medieval millenarians to modern beats and drop-outs, nothing has in this respect changed, even after the emergence of modernity's 'mass literate' societies (so called), where so much investment and effort is put into each particular form of politicised mass schooling. All that has happened is that it is more difficult to be defined as radically 'other'. Hence the 'threat' once posed by the nomads and anarchists is now posed by jihadists and drop-outs.

In the spirit of seeking to understand origins as geneses and, following this line of argument, as geneses which entail modes of invention, I wish to offer a way

3 Cf. Mahmoud Ezzamel and Keith Hoskin, 'Retheorizing Accounting, Writing and Money, with Evidence from Mesopotamia and Ancient Egypt', *Critical Perspectives in Accounting*, 13, 3 (2002), 333–67.

4 Gilles Deleuze and Felix Guattari, *A Thousand Plateaus: Capitalism and Schizophrenia* (London: Athlone Press, 1988), pp. 351–423.

of understanding who we are and how we 'become' which will draw on but then perhaps go beyond Foucault's focus on certain historical moments of discontinuity as particularly fertile moments of genesis in terms of how we think, write and act. I shall focus on just one such moment, which Foucault first brought to attention in a new way in his archaeological work, *The Order of Things*. It is a genesis/invention which took place in the late eighteenth and early nineteenth centuries, and which gave birth to what we might describe as a new knowledge 'ecosystem', which I would characterise as that of the 'knowledge disciplines'.

I do so here because that genesis moment can also be read, thanks to Foucault's equally groundbreaking insights in his first great 'genealogical' work, *Discipline and Punish*, as the birth of a new power 'ecosystem', that of what many, following Foucault, have described as 'disciplinary power'. What I therefore seek to do is illustrate, in the way that can be done for the inventions of accounting or writing, how there was a genesis process which led not just to new forms of knowledge and power, but also to new modes of pursuing and exercising both. In this way the Foucauldian hybrid 'power/knowledge' can be given a more precise designation when applied to characterising and understanding our contemporary era, which can, I suggest, be named as a 'doubly disciplinary' era. For the disciplinary, sub-disciplinary and trans-disciplinary knowledge generated in the modern knowledge ecosystem – the ecosystem within which we all, who write as academics and intellectuals, write today – interplays with and indeed enables the exercise of the modern forms of disciplining selves and populations. Discipline begets discipline: a twinning of terms whose writ runs simultaneously in both directions.

What I propose, briefly, to do is as follows. First I propose to summarise what Foucault has to say in *The Order of Things* about the transformation of what he calls an older knowledge *episteme*, which is in his terms 'representational', to a new and very different one, which I characterise as 'disciplinary'. Then I shall suggest how this transformation, which he describes there in archaeological terms, can also be described in his later 'genealogical' terms, and can thereby provide an explanation of the 'how' and 'why' of this genesis moment which is not available under the archaeological gaze. I then want finally to suggest that, if we accept that this is the genesis of a 'doubly disciplinary' era, we may then get a new understanding of certain significant developments of that era as 'inventions': the particular development which I shall suggest may be understood in that way is 'management', that breakthrough approach to getting people to get things done whose early history was first adequately traced by Alfred Chandler in *The Visible Hand* (1977).

Here I seek to amplify Chandler's insight that the 'visible hand' of management first materializes in the early to mid nineteenth century in the USA and then made possible the modern economic world dominated by multinational and transnational corporations, wherein the 'invisible hand' of the market is subordinated to the oligopoly power deployed by these managerial behemoths.[5] Details of Chandler's interpretation of the triumph since 1900 of the multi-divisional or M-form structure have been challenged, but his explanation of the initial breakthrough as a response to some form of new economic demand has been less challenged.[6] I suggest here that any demand-response explanation fails to capture how management was one mode of invention ensuing from how humans had begun to think, act and write in a doubly disciplinary era.

II. On the Genesis of Double Disciplinarity: the 'Archaeological' Study of a Knowledge Shift

In the second half of *The Order of Things* Foucault focuses on what got said (and so implicitly remained unsaid) in the seventeenth and eighteenth centuries in three particular fields of knowledge discourse – those concerned with words, living things and value. He then contrasts this range of things said and unsaid with the very different range of things that materializes within each field, as new and radically different successor discourses get articulated from the late eighteenth century on.

In each era he sees certain (and systematically different) regularities as to what constitute the 'objects of knowledge' in each of these three fields and as to how those objects get talked and written about, to the point where he can see a common way of knowing, or 'episteme', as shaping both what gets known and the limits of the knowable. In the seventeenth and eighteenth centuries he sees the 'episteme' as 'representational', as the knowledge discourses through which words, living things and value are studied operate through the writing of summative texts incorporating tableaux and charts of essential characteristics. Thus words, living things and value are studied, named and arrayed as essentially spatially representable and so fixed, frozen in time, or *sub specie aeternitatis*.

5 The recent credit crunch does not negate the oligopoly power of transnational firms. Indeed from the 'visible hand' perspective the collapse of financial 'markets' is a direct consequence of US banking and finance, once retail banks were allowed to operate beyond regional markets in the 1980s, having adapted the classic form of the managerially run oligopolistic venture to the business of speculating in ever-expanding capital markets.

6 However, see Keith Hoskin and Richard Macve, 'The Genesis of Accountability: The West Point Connections', *Accounting, Organizations and Society*, 13, 1 (1988), 37 – 73, and 'Knowing More as Knowing Less? Alternative Histories of Cost and Management Accounting in the USA and UK', *Accounting Historians Journal*, 27, 1 (2000), 91 – 149, and the historical debates reviewed there.

More specifically, the study of words was focussed on a 'general' or 'universal grammar', which would define the characteristics of all languages: that of things on 'natural history', the identification and representation in tableaux of genera and species; and that of values on the analysis of a wealth defined by the supposedly fixed store of silver and gold reposing in the Earth, the spatial horizon and limit of monetary value.

Then, late in the eighteenth and early in the nineteenth centuries, he charts how, in a radical discontinuity, new discourses suddenly get articulated in these three areas, in each of which there are now new regularities as to what constitute the 'objects of knowledge' and to how they get talked and written about. So the study of words engaged in a new object of study, specific languages in specific times and places, undertaken via a new discourse of 'philology'. The study of living things engaged for its object of study with the temporal entity, the 'organism', whose 'organisation' was made up of 'functions' (digestion, excretion, reproduction, etc), and was studied through dissection in a new discourse of 'biology'. Finally the study of values engaged for its object of study in tracking processes of exchange and production and finance and government within specific societies, and studied these through the amassment and analysis of data on such categories, in a new discourse of 'political economy', whose godchild would be 'economics'.

Here there are not simply different 'objects of knowledge' through which words, living things and value are now studied, there is (as part of the explanation of why this difference should manifest itself) a different *epistemic assumption*. The episteme can no longer be understood as one where the focus is 'representational'. None of these objects of knowledge turns out to be purely spatially representable in the sense of being assumed to be fixed, frozen in time, or *sub specie aeternitatis*. Necessarily, as an indispensable *semiotic device*, spatial representation remains in play – as can be seen in the texts and tropes of virtually all of the modern knowledge disciplines, and certainly in all those of the natural and human 'sciences'. But its role is now to depict and capture temporality, in the variously *emergent* and *changing* forms of language, organism and monetary value. These have become knowledge objects which are studied in an epistemological frame where they can no longer even be *assumed* to be fixed, frozen in time, or seemingly eternal. Instead *temporality* is now the essential precondition assumed and already taken for granted in the very moment or event of *beginning* to study each of these knowledge objects, despite their continuing radical difference from each other.

Each of them will henceforward be studied, understood and extended along its own particular methodological and narrative trajectories via the processes of research. But all will continue to be known, as they have been down to today (and will continue to be until a further epistemic assumption displaces that under whose sign we all think and write), via a contested series of continuities and/or discontinuities occasioned by their *necessary temporality.* Finally these knowledge objects become 'known' by us as a different form of knowledgeable subject: human subjects who are ourselves necessarily temporally situated subjects, who no longer generate truths that will stand *sub specie aeternitatis*, but 'disciplinary' truths which, insofar as they get accepted by the exponents of given disciplinary discourses in a given time and space, are established only to be displaced and re-written in a potentially infinite series of newer and deeper insights, within and beyond particular 'paradigms'.

Foucault's shorthand to describe this change in assumption is to say that 'man' (in the sense of 'the human', or '*l'homme*') entered the field of knowledge discourse in a new way, which consisted in man becoming simultaneously the object of study as well as the subject doing the studying. The consequences of that have been with us in all the great dualist variants of the past two centuries, absolutism/relativism, objectivism/subjectivism, and not excluding that 'critical' re-take on medieval categories, 'critical' realism/nominalism. At the same time, Foucault signals something for us as contemporary knowing experts engaging in producing visible knowledge, that we are the inheritors and carriers, in our modern *episteme*, of a mode of knowing never previously articulated by humans, but now seemingly unavoidable, because of how we come to know (before we even know it) that there could not be any other way of 'coming to know' than that we (as humans) are born and raised into.

III. On the Genesis of Double Disciplinarity: the Genealogical Study of a Power (and Knowledge) Shift

The articulation of this theme in *The Order of Things* is undertaken, as noted already, 'archaeologically', in the sense of comparing what is visible (as text and discourse) in different strata. There is a discontinuity, a rupture, a silencing of the old in favour of the new which is simply not detectable in or deducible from what was said, written and thought before.

Of course it would be disingenuous to suggest or imply indeed that the old discourses simply disappear or have no effect thereafter; for they can continue to animate significant theorising down to today. One can note, for instance, a continuance of the principle of 'general grammar' in the linguistics approach of Chomsky, who has long acknowledged how the attempts of the seventeenth-century Bishop Wilkins to establish a 'universal grammar' inspired his own attempts to articulate a 'generative grammar', attempts which have led in turn to studies such as that into a 'generative semantics'. Similarly the tableaux of genera and species inform the Darwin and Spencer studies of evolution. But in each case temporality is now embedded, as is the concern with temporal man as object as well as subject of study. So the very term 'evolution' displaces spatial fixity in favour of temporality as an integral given within and across 'species', and an implicit recognition that the human subject must necessarily have its own species as object within a temporally emergent continuum of all species. Generative grammar (or semantics) similarly set themselves up as quasi-Platonic categorisations of 'languaging' which can only be revealed or known to be there through or at the temporal surface of specific languages, discourses, speakings and scripts.

However, Foucault then launches an approach, in his genealogical works, that engages in a different way with the 'how' and 'why' of change by bringing a 'genesis' focus into play, in considering the development (both individual and social) of us (as) 'humans'. This was not because he saw origin as an empty or meaningless descriptor of human events, nor because 'originating' has no meaning as a form of human practice. But he was instead, so it seems to me, concerned in his consequent new guise as 'a historian of the present' to promote a continually thoughtful reflection on the human process of *becoming*. Building on one of the great prior reflections of the twentieth century – Heidegger's dwelling on how it is that human beings are always carriers of 'being' but only as beings which are continually 'becoming' – he arguably devoted his own work to making clear, in new ways, how our pasts structure how we as humans think, act and live our lives at a particular space and time.

In *Discipline and Punish*, Foucault traces such a process at both individual and social levels, particularly through a focus on certain practices and how they engage us both within the individual subject and at the level of the population of subjects. In the central section of the book, 'The Means of Correct Training', [7] he focuses in particular on practices or technologies of 'hierarchical observation' (pp. 170 – 77) and 'normalising judgement' (pp. 177 – 84), and then argues that these two sets of technology come together in a new practice, 'The Examination' (pp. 184 – 94), a practice that, combining both, is then exercised in classrooms on students, turning them into 'a whole field of knowledge' (p. 186), namely the knowledge transcribed into that new and now ubiquitous sub-literary form, the textbook, which becomes the object of the now also ubiquitous formal examinations undergone by children (and now also adults) across the human world. From this new population of examinees there emerge, from among all those ex-examinees, successive and more numerous generations of *examiners*. Most of these focus on producing the succeeding waves of textbooks and tests that

7 Michel Foucault, 'The Means of Correct Training', in *Discipline and Punish* (London: Allen Lane, 1977), pp. 170 – 94. Further references are given after quotations in the text.

students must undergo; but an increasing number of the most successful examinees, those most adept in the many and various knowledge disciplines that begin to emerge, re-make themselves as knowledge-*producing* subjects, who *deploy* examination to study objects and questions of concern within such disciplines and so, as individuals, produce new contributions to disciplinary knowledge, while *en masse* they contribute to the continuous expansion and proliferation of the project of knowledge disciplinarity.

So we enter a new age of man (*l'homme*), where the principle of what it is to be an individual is reversed. Where in earlier societies 'the more one possesses power or privilege, the more one is marked as an individual' (p. 192), so that individualisation can be seen as 'ascending', now the practice of examination, undertaken by knowledgeable ex-examinees, makes each individual a 'case', so that 'individualization is 'descending': as power becomes more anonymous and more functional, those on whom it is exercised tend to become more strongly individualized' (p. 193). This is, of course, a classic and much cited example of how 'disciplinary power' is exercised to produce 'docile bodies'. But the exercise of that power, as Foucault then himself points out, only proceeds and maintains itself through the intervention of experts schooled in the relevant knowledge disciplines. As he observes: 'all the sciences, analyses or practices employing the root 'psycho-' have their origin in this historical reversal of the procedures of individualisation' (p. 193).

In earlier work, [8] I have suggested that Foucault here has a devastatingly important insight about a change in how we as humans learn, so as to become subjects who both live under, but also learn how to exercise, constant and increasingly ubiquitous forms of 'hierarchical surveillance' and 'normalising judgement'. At the same time, it is necessary to specify more precisely what is new about the practice of examination. For formal examination as such had been familiar to students in European pedagogic settings from the end of the twelfth century. Formal and ritualised examination by teachers, allied to the right ceded to bodies of teachers by higher authorities (initially the Catholic Church) to give licenses or degrees on the basis of performance in such examination, had been the central practice constituting those institutions that survive and flourish till today, the universities. What was new was the supplementing of the old practice of formal examination, which remained *oral* down into the eighteenth century, by two new practices. The first was getting the students to write what would be examined, and the second was putting specific numerical marks or grades on each performance, where previously the practice had been one of differentiated

8 Keith Hoskin, 'Foucault Under Examination: The Crypto-Educationist Unmasked', in *Foucault and Education*, ed. by Stephen Ball (London: Routledge, 1990), pp. 29–53.

qualitative evaluations (e.g. excellent, good, acceptable, not acceptable, without merit).

What was new then in terms of pedagogic practice was a tripartite unity: getting students to write, examining them on their writing as well as orally, and giving numerical grades for each and every 'learning performance'. This was a precondition for 'normalising judgement' in the sense of constructing numerical norms for given sets of performances and of distributions from top to bottom. It was also the precondition for being able, at the level of the individual human subject, to give an objective measure of performance (e.g. 10 out of 10, or 0 out of 10) as well as an ordinal ranking from first to last (a practice already developed by the sixteenth century, by the Jesuits among others). The principle of competition could be brought into play across the whole of the relevant population being examined, as also could the principle of objective and introjected 'success' and 'failure', constantly redefined by your latest performance and stretching out into the future, as a sword of Damocles defining who as a subject you 'are' at each given time and place. An earlier extensive human concern with 'great individuals', i.e. 'memorable selves', gets supplemented with a new writing, examining and grading of 'calculable selves', who can be summed into and compared as different 'populations' of such selves. And one may note, as 'calculable selves', we also become 'calculating selves', both in the sense that what follows is the first human era not only of mass literacy but mass *numeracy*,[9] and in the sense that our era runs things through numbers and proliferates precisely calculating forms of Machiavellianism.

Once such an amendment of Foucault's initial insight about the power of 'examination' is made, it then becomes possible to revisit the archaeological insights of *The Order of Things* using a *genealogical* approach. What then becomes apparent[10] is that those who participate in the genesis of the new knowledge discourses identified by Foucault, philology for words, biology for living things, and political economy for value, all experience the first articulations of the new pedagogic approach which combines writing, examining and grading. Interestingly, each of the specific geneses takes place independently, in the sense that each knowledge rupture or breakthrough takes place in a different country, through a different initial application or combination of the new practices, and in a different institutional setting. But all occur over the same decades, between the 1760s and 1780s, and within each field new objects of knowledge are constituted in which *temporality* is integral; and from each

9 Patricia Cline Cohen, *A Calculating People* (Chicago: University of Chicago Press, 1985).
10 Keith Hoskin, 'Education and the Genesis of Disciplinarity: The Unexpected Reversal', in *Knowledges: Historical and Critical Studies in Disciplinarity*, ed. by Ellen Messer-Davidow and others (Charlottesville: University of Virginia Press, 1993), pp. 271–304.

genesis there quickly flows a whole series of new knowledge disciplines, a series which expands and proliferates around each object of knowledge down to today, on the model of what happened in the field of individualization with the genesis of all those 'psycho-' disciplines.

So in Germany, the first great student and exponent of philology is usually taken as Friedrich Wolf, who established in his Philological Seminar at Halle in 1786 the practice of hermeneutic and detailed textual criticism of series of texts to fix them in their time, genre and authorship. Yet we now know [11] that this practice was begun, in the new space of the seminar room, in 1763 at Göttingen by Christian Heyne. More specifically Heyne introduced the practice, within the weekly seminar, of students having to write a critical essay, or *Aufsatz*, on a philological topic which was then subject to joint examination by the class and teacher. Further, Wolf enrolled 'in Heyne's Philological Seminar in 1777 as the first official student of philology'. [12] So Wolf, having learned under this form of the new disciplinary pedagogy, becomes the first great articulator of philology as the new knowledge discourse of words; and hereafter there develops the whole disciplinary study of words in all aspects, from the most 'psycho-' form of linguistics to the most poststructuralist study of literature.

Meanwhile in France, the constitution of the study of the living organism's physiology, in the new discourse of 'biology', is usually associated with Bichat and his commitment to the constant examination of specimens, particularly via 'dissection'. However, as Richard Jones shows, [13] Bichat learned his laboratory-based scientific practice as a student of the surgeon Pierre Dessault, who, in the now-familiar doubly disciplinary way, required students to develop their specialist expertise in investigating living things through a constant system of pedagogic examination. Every ten days the students were required to undergo rigorous examinations on material covered up to then, examinations which included both text-based and practical laboratory tests.

More generally, [14] this new active engagement with practical forms of investigative 'examination' in laboratories can be traced across the fields of scientific endeavour in France from the 1770s, with the key figure being Gaspard Monge, who introduced this practice as a teacher in the Royal College of Artillery at Mezières. That mattered because Monge became, after the Revolution, the key figure in scientific knowledge production and pedagogy, as central player in the

11 Ibid., pp. 281–83, and Robert Leventhal, 'The Emergence of Philological Discourse in the German States, 1770–1810', *Isis*, 77 (1986), pp. 243–60.
12 'The Emergence of Philological Discourse', p. 246.
13 Richard Jones, 'Educational Practices and Scientific Knowledge: a Genealogical Reinterpretation of the Emergence of Physiology in Post-revolutionary France', in *Foucault and Education*, pp. 78–104.
14 'Education and the Genesis of Disciplinarity', pp. 285–88.

setting up of the *École Polytechnique* and the other *Grandes Écoles* and as Napoleon's most trusted and senior scientific officer. In this instance, what had been a passive and non-examined way of studying science (through attendance at professorial demonstrations and lectures) became an active and examining engagement with the relevant physical, chemical and physiological objects. For a short period, as one observer has put it: 'The French community of science predominated in the world to a degree that no other national complex has since done or ever had done'. [15] As it did so, the new active, examining, writing and grading (disciplinary) approach to doing science took root, then to be disseminated across Europe and the US, and then the world, as 'science as such'.

Finally in Scotland, a slightly different but still structurally similar pedagogic shift led to the constitution of political economy. Here at the University of Glasgow a further new pedagogic space was developed, analogous to the seminar and the laboratory: the classroom. [16] David Hamilton [17] has established that the classroom as a space for simultaneous instruction of 'classes' really takes off in the early nineteenth century. But the first English-language usage of the term he has traced is at a Glasgow University faculty meeting in 1762, where it was decided to transform a college chamber into a 'class room'. Again there was specific pedagogic intent, similar to that with the seminar: but in this case students were required to study a text in advance of attendance at the 'class room', and would then be subject to extempore examination of each individual in turn, with no pre-knowledge of what questions each would face. This was designed to promote a group dynamic of 'sympathy' and an individual one of 'emulation' and was popularised at Glasgow by the professor of logic from 1774, George Jardine. But Jardine's teacher, in the 'class room' setting, was Adam Smith.

Here the pattern of genesis differs slightly: for if Jardine is Wolf to Smith's Heyne, yet it is Smith who brings the term and practice of 'examination' into a historically unprecedented role as the means to the genesis of a new *philosophical* construct of the human subject as 'self'. This takes place in that work whose first publication is currently being celebrated since it took place 250 years ago: *A Theory of Moral Sentiments* (1759). The work is increasingly interpreted (even if beyond the purview of most economists) as Smith's great intellectual contribution, with *The Wealth of Nations* (1776) simply one application of its radical re-thinking of the nature of the human subject. But in fact the great re-thinking in *Theory* does not occur in the first edition but only in the second

15 Charles Coulston Gillispie, *Science and Polity in France. The End of the Old Regime* (Princeton: Princeton University Press, 1980), p. 9.
16 'Education and the Genesis of Disciplinarity', pp. 288–95.
17 David Hamilton, *Towards a Theory of Schooling* (London: Falmer Press, 1989), pp. 75–78.

edition of 1762. Here Smith solves the Hobbesian problem of how a rational and just society can be built given the innate selfishness of humans as 'selves'. Various Enlightenment thinkers, including Hume, had attempted to refute Hobbes's conclusion that the Leviathan of a society made up of such selfish selves must have an authoritarian ruler. Smith, in the second edition of *Theory*, is the first to come up with a formulation of the self that offers a philosophically adequate (if itself newly problematic) refutation of Hobbes. As Milton L. Myers has pointed out in *The Soul of Economic Man* (1983), [18] previous philosophers had argued that the self did have constraints against the selfishness of selves defined by Hobbes, but these were all external to the self. So it had been argued that the self was subject to being moved by some independent touchstone of ethicality, such as considering how 'mankind' or 'an impartial spectator' (Hume's particular suggestion) would view an action. But these all failed to overcome the objection that the self would in the last analysis reject such external influences, and so continue to put its own selfish interests first.

In the first edition of 1759, Smith retains but extends Hume's idea that an 'impartial spectator' solves the selfishness problem, by introjecting this spectator *within* the self, so that there are now two *personae* inside us, one active and the other passive. So he proposes at the opening of Book 3 of *Theory* that 'we must imagine ourselves not the actors but the spectators of our own character and conduct'. But Smith then received a letter from a reader objecting that such a passive spectator constituted no *necessary* independent touchstone of ethicality, for 'if conscience is a reflection of social attitudes, how can it ever differ from or be thought superior to popular opinion?' Smith responded in the second edition by re-writing the passage in a way that turned the essentially passive spectator *persona* into an independently *active* one. The choice of term for this new *persona* is instructive, if by this point in this analysis predictable. For now the self is constructed as being able to overcome the selfishness problem through the incorporation within it of an active second *persona* which is ethically independent of the first. As Smith puts it:

> When I endeavour to examine my own conduct [...] it is evident that, in all such cases, I divide myself as it were into two persons; and that I, the examiner and judge, represent a different character from that other I, the person whose conduct is examined into and judged of. The first is the Spectator whose sentiments with regard to my own conduct I endeavour to enter into [...]. The second is the Agent, the person whom I properly call myself. [19]

18 Milton L. Myers, *The Soul of Modern Economic Man. Ideas of Self-Interest: Thomas Hobbes to Adam Smith* (Chicago – London, University of Chicago Press, 1983).

19 Adam Smith, *The Theory of Moral Sentiments*, ed. by David Daiches Raphael and Alec Lawrence MacFie (Oxford: Clarendon Press, 1976, orig. pub. 1759), Book 3, Chapter 1, pa-

Now of course it is not 'evident' at all, before Smith wrote it, that one can divide one's self in that way, still less that one can then claim that one of these two is more real, as 'the person whom I properly call myself', than the other. Indeed the examining/judging persona may be equally argued to be 'more real'. But regardless of that, the philosophical outcome is the genesis of the human subject as a new form of doubled and doubly active self, of a kind without apparent precedent in the previous history of philosophical reflection on the self. At the same time, it is a form of self that many of us contemporary human subjects may recognise as or in our 'selves', and as a feature of ways of modern self-understanding from 'psycho-' texts to the Diary of Bridget Jones.

But perhaps its most significant translation has occurred not in psychology at all but in making possible what Smith is most famous for, the genesis of the seductive dangers of modern economics. For a self that is simultaneously spectator-examiner and agent-examinee can perform the previously inconceivable and prestidigitatory feat of engaging in *rational self-interest*. Thus another genesis becomes possible, as the title of Myers' book so aptly puts it: the 'soul' of a new kind of man, *homo economicus*, who can, once Smith has written his second great work, be assumed to inhabit a world of self-clearing and equilibrating markets as the central player in the modern disciplinary tragicomedy, economics.

So we have three disparate knowledge fields where, over the same decades in the late eighteenth century, new modes of pedagogy are introduced in three different countries and three different institutional settings, the seminar, the laboratory and the classroom. But once we realise this, it becomes possible to undertake a genealogical analysis of the fields studied archaeologically in *The Order of Things*, and to begin to see how the 'genesis' of each and all of the new disciplinary knowledge fields occurs when humans begin to learn under the doubly disciplinary pedagogic practices under which no humans had learned before.

Once they begin to live out the new pedagogic truth that 'real learning' is that which takes place under constant writing, examining and grading, they can progress, as ex-examinees, to articulate and develop the new and parallel epistemological truth, that 'real knowledge' is that produced through the self-same practices. Thus is launched the virtuous/vicious circle of self-justification for all things disciplinary, whether manifestations of disciplinary knowledge, power or some amalgam of the two; and as a corollary, today learning under writing, examining and grading has become so taken for granted by and in us, that it is

ragraph 2, pp. 110–11. The objection came in a letter from Sir Gilbert Elliot, and is discussed by Raphael and MacFie in their *Introduction*, p. 16.

seldom even noticed how strange, unprecedented and even bizarre such learning is.

IV. From Double Disciplinarity to the Invention of Management

In a series of works with Richard Macve, [20] I have pursued some significant developments which follow on from the genesis of world populated by disciplined and self-disciplining selves who seek to become, via systems of mass education which now extend to the university level, experts in the multifarious and ever-shifting forms of modern disciplinary knowledge. In particular we have sought to re-interpret the development of management, as traced by Chandler, [21] as a doubly disciplinary breakthrough, through showing, in a way similar to that undertaken in the last section, that there are a few key people who are among the first to undergo the new pedagogic approach entailing constant writing, examining and grading who then translate the knowledge disciplinarity as well as the disciplining of action and thought to business contexts.

In particular we have argued that there is a genealogical path from the *École Polytechnique* to the US Military Academy at West Point to the first business enterprises which Chandler identifies as developing the first successful forms of modern business management. For the pedagogic practices, plus the disciplinary knowledge texts, brought into use, under Monge's *aegis*, at the *École Polytechnique* are imported in 1815 into the US by the man who would in 1817 become West Point's third Superintendent, Sylvanus Thayer. [22] As West Point Superintendent, Thayer introduced the first pedagogic regime based on constant writing, examining and grading in the US, and produced graduates who were at

20 Keith Hoskin and Richard Macve, 'Reappraising the Genesis of Managerialism: A Re-examination of the Role of Accounting at the Springfield Armory, 1815–1845', *Accounting, Auditing and Accountability Journal*, 7, 2 (1994), 2–29; 'The Genesis of Accountability'; Keith Hoskin and Richard Macve, 'The Western Railroad Crash of 1841: the first Manageriogenic Disaster?', paper presented at the *Interdisciplinary Perspectives on Accounting* Conference, July 2009.

21 A. D. Chandler, *The Visible Hand* (Cambridge, MA: Harvard University Press, 1977).

22 This is because Thayer was sent by the US government to Europe to identify and bring back the best modern educational texts and systems for military training. He decided (quite rightly, given Gillispie's observation) that the mathematical and engineering texts and the highly intensive examining pedagogy used to teach them in classes differentiated by 'ability' as tracked by exam performance were by far the best available. He also had the good fortune to be in Paris in the summer after the Battle of Waterloo and so was able to purchase both the texts and one of the teachers from the *École Polytechnique* – something that would probably have been impossible otherwise. One consequence was that for the first several years of his Superintendship, the students devoted the whole of their first year to the study of mathematics and French, the latter an essential since the textbooks had not yet been translated.

that time by far the finest engineers and surveyors there, because of their exposure to the then-unparalleled scientific and mathematical knowledge developed and turned into textbook form by the leading French academics.

What we have then traced is how the first breakthroughs into management – identified by Chandler as taking place at the US Armory at Springfield, where as he sees it 'single-unit' or factory management was developed, and on the US railroads, where 'multi-unit management' was developed – are the result of structural and processual innovations in administrative coordination introduced by graduates of Thayer's West Point.

In particular at the Springfield Armory, an 1819 graduate, Daniel Tyler, undertakes in 1831 a 6-month study of each of the almost two hundred activities involved in the manufacture of a musket and comes up with a *standard* time that the production of each piece in the musket *ought* to take. He does this by examining 'watch in hand' each activity and recording how long it does take and then recalculating how long it should once inefficiency and what F. W. Taylor would call 'soldiering' were factored out. He then instituted an efficient flow system of production through each shop, which incorporated an existing system for minimising wastage of materials alongside the new standard-based form of labour-time accountability he had devised.

A number of people make various kinds of attempt at similar administrative coordination on the US railroads, including one, George Whistler, who is a graduate in the 1819 West Point class alongside Tyler. We now see Whistler as demonstrating how not to construct a successful multi-unit management system. [23] However the man who first succeeds in constructing a line-and-staff management structure with a constant processual system for tracking performance and outcomes and then devising strategies and targets via the constant collection, examination and evaluation of detailed accounting and statistical information is a West Pointer of a slightly later generation, Herman Haupt, who succeeds in bringing structure and process together into the first recognisable form of multi-unit and divisionalised business on the Pennsylvania Railroad between 1849 and 1856. [24]

23 Whistler has been credited by Chandler with developing a successful staff-based central management function with functional departments reporting into it. Our recent reanalysis shows that, while he does operate with departments and has a central management function, it is actually defective. Whistler's failure to have a failsafe system for relaying commands leads directly, in October 1841, to the first fatal head-on collision of trains on a US (single-track) railroad. If anything, on the analogy of 'iatrogenic disease', Whistler launches the first (but by no means the last) 'manageriogenic disaster'.

24 Keith Hoskin and Richard Macve, 'The Pennsylvania Railroad, 1849, and the Invention of Modern Management', paper presented at the American Accounting Association Annual Meeting, Orlando (FL), August 2004.

V. Conclusion: The Disciplinary Genesis of Management as 'Invention'

My purpose here has not been to rehearse the detail either of these breakthroughs or of the historiographical debate they have occasioned. [25] Rather it is to speculate if only briefly on how this genealogical tracing of a change in how humans learned – and indeed learned to learn – which passed from European elite educational institutions to West Point (and then on to modern US mass education) can be seen as leading to management as an 'invention'.

One thesis of my work with Richard Macve is that management has been constantly mis-read and misunderstood as being a business (and power) breakthrough (which leaves the study of management to date as something inadequately theorised). But businessmen, as the detailed history we have reviewed makes clear, were perhaps the last people to be likely to come up with the new combination of structuring and processing imposed on workplace and business activity at the Springfield Armory and on the Pennsylvania Railroad. The men who did construct modern management come from a different epistemic world, to use Foucault's archaeological term. It is their double grounding in mathematico-scientific disciplinary knowledge expertise combined with a disciplinary learning experience permeated with constant writing, examining and grading which enables them to solve an age-old human problem, that of 'getting people to get things done', in a wholly new and unprecedented discursive way.

For they do not only solve the problem of coordinating the action of workforces in behaviourally disciplined and knowledge-discipline based ways – although this is what is essential to enable the visible hand of management to dominate, economically, the invisible hand of the market (and so consign all forms of naïve market-based economic theorising to their continuing tragicomic status). They also launch the discursive possibility of all the subsequent knowledge disciplines and sub-disciplines which now constitute the study and doing of business: accounting (financial and managerial), human resource management, marketing, operations management, strategy, and so on.

This breakthrough, we would now argue, constitutes an invention, in just the way that other great breakthroughs of the same era are seen as doing, for instance steam power and telecommunications. Just as Watt's invention of the efficient steam engine transformed the potential of machine power sources and launched the machine age, and Morse's effective telegraph system transformed the potential for telecommunications and launched the telecoms age, manage-

25 'Knowing More as Knowing Less?'; 'The Western Railroad Crash of 1841: the first Manageriogenic Disaster?'.

ment as invention transformed the potential for 'getting people to get things done' – the great human problem of 'indirection' – and so launched the age of the giant corporation and the managerial state, and indeed all the myriad forms of modern self-management.

In that respect, I suggest, management's success is like that of the steam engine and the telegraph, in the sense that it not only generated a new solution to an ancient problem, it did so in such a profoundly influential and powerful way that it transformed the basis on how we humans would henceforward live our lives. As such it was an invention which follows on from a more profound moment of genesis: a moment which can equally be interpreted as an origin, and that is a term I would not wish to preclude. All I am seeking, in that respect, to preserve is a sense of the integral necessity of the temporal to such transitions, so that there is always something already there which is the precondition for what displaces it, while never anticipating specifically or in a clearly foreseeable way what ensues. And alongside that, I am seeking to allow for the mystery of how what was not said before now becomes insistently said, along with a widely disseminated sense first that the product of genesis had always been said and second that it will always continue to be: two ways of seeing human being/becoming which are equally misleading and untrue.

Bibliography Part II

Abel, M. H., *An Empirical Reflection on the Smile* (Lewinston: Edwin Mellen Press, 2002).
Baker, M. C., *The Atoms of Language* (New York: Basic Books, 2003).
Balda, R. P., I. M. Pepperberg, and A. C. Kamil, eds, *Animal Cognition in Nature* (San Diego: Academic Press, 1998).
Ball, S., ed., *Foucault and Education* (London: Routledge, 1990).
Bartocci, M., J. Winberg, G. Papendieck, T. Mustica, G. Serra, and H. Lagercrantz, 'Cerebral Hemodynamic Response to Unpleasant Odors in the Preterm Newborn Measured by Near-Infrared Spectroscopy', *Pediatric Research*, 50, 3 (2001), 324 – 30.
Bickerton, D., *Language and Species* (Chicago: University of Chicago Press, 1990).
Botha, R., 'Pidgin Languages as a Putative Window on Language Evolution', *Language and Communication*, 26 (2006), 1 – 14.
Brazelton, T. B., *Clinics in Developmental Medicine, No. 88: Neonatal Behavioral Assessment Scale (NBAS)*, 2nd edn (Philadelphia: Lippincott, 1983).
Brosnan, S. F., and F. B. M. de Waal, 'A Proximate Perspective on Reciprocal Altruism', *Human Nature*, 13 (2002), 129 – 52.
Burkart, J. M., E. Fehr, C. Efferson, and C. P. van Schaik, 'Other-regarding preferences in a non-human primate: Common marmosets provision food altruistically', *Proceedings National Academy Science*, 104 (2007), 19762 – 66.
Chandler, A. D., *The Visible Hand* (Cambridge, MA: Harvard University Press, 1977).
Cheney, D. L., 'The acquisition of rank and the development of reciprocal alliances among free-ranging immature baboons', *Behavioural Ecology & Sociobiology*, 2 (1977), 303 – 18.
Chomsky, N., 'Review of B. F. Skinner's Verbal Behaviour', *Language*, 35, 1 (1959), 26 – 58.
Chomsky, N., *Rules and Representations* (New York: Columbia University Press, 1980).
Chomsky, N., *Lectures on Government and Binding* (Dordrecht: Foris, 1981).
Chomsky, N., *Language and Problems of Knowledge: the Managua Lectures* (Cambridge, MA: MIT Press, 1988).
Chomsky, N., *Knowledge of Language: Its Nature, Origin and Use* (New York: Praeger, 1989).
Chomsky, N., 'Three Factors in Language Design', *Linguistic Inquiry*, 36, 1 (2005), 1 – 22.
Chomsky, N., 'Universals of Human Nature', *Psychotherapy and Psychosomatics*, 74 (2005), 263 – 68.

Christiansen, M. H., and S. Kirby, eds, *Language Evolution. The States of the Art* (Oxford: Oxford University Press, 2003).

Clutton-Brock, T. H., P. N. Brotherton, M. J. O'Riain, A. S. Griffin, D. Gaynor, L. Sharpe, R. Kansky, M. B. Manser, and G. M. McIlrath, 'Individual Contributions to Babysitting in a Cooperative Mongoose, *Suricata Suricatta*', *Proceedings of the Royal Society of London*, 267 (2000), 301 – 05.

Cohn, J. F., A. J. Zlochower, J. Lien, and T. Kanade, 'Automated Face Analysis by Feature Point Tracking Has High Concurrent Validity with Manual FACS Coding', *Psychophysiology*, 36, 1 (1999), 35 – 43.

Cohen, P. C., *A Calculating People* (Chicago: University of Chicago Press, 1985).

Cole, J. K., ed., *Nebraska Symposium on Motivation 1971* (Lincoln: University of Nebraska Press, 1972).

Collins, W. A., ed., *Minnesota Symposium on Child Development*, vol. 11 (Hillsdale: Lawrence Erlbaum, 1978).

Culicover, P. W., and R. Jackendoff, *Simpler Syntax* (Oxford: Oxford University Press, 2005).

Culicover, P. W., and R. Jackendoff, 'The Simpler Syntax Hypothesis', *Trends in Cognitive Science*, 10, 9 (2006), 414 – 18.

Darwin, C., *On the Origin of Species by Means of Natural Selection, or the Preservation of Favoured Races in the Struggle for Life* (London: John Murray, 1859).

Darwin, C., *The Descent of Man and Selection in Relation to Sex* (London: John Murray, 1871).

Darwin, C., *The Expression of the Emotions in Man and Animals* (London: John Murray, 1872).

Deleuze, G., and F. Guattari, *A Thousand Plateaus: Capitalism and Schizophrenia* (London: Athlone Press, 1988).

Dondi, M., A. Costabile, M. Rabissoni, C. Gianfranchi, O. Lombardi, and C. Corchia, 'Il sorriso del neonato prematuro: una ricerca esplorativa' ('Smiles of Premature Neonates: An Exploratory Investigation'), *Giornale Italiano di Psicologia*, 31 (2004), 377 – 99.

Dondi, M., D. S. Messinger, M. Colle, A. Tabasso, F. Simion, B. Dalla Barba, and A. Fogel, 'A New Perspective on Neonatal Smiling: Differences Between the Judgments of Expert Coders and Naïve Observers', *Infancy*, 12, 3 (2007), 235 – 55.

Dondi, M., A. Costabile, T. Vacca, L. Franchin, S. Agnoli, O. Lombardi, and C. Corchia, 'Alle origini delle emozioni: uno studio pilota sul sorriso e sull'espressione facciale di distress nel neonato pretermine di età gestazionale molto bassa' ('At the Origins of Emotions: A Pilot Study on the Facial Expression of Distress and Smiling in Very Low Gestational Age Preterm Newborns'), *Psicologia Clinica dello Sviluppo*, 12, 1 (2008), 89 – 109.

Duchenne (de Boulogne), G. B., *Mécanisme de la physionomie humaine ou analyse électrophysiologique de l'expression des passions* (Paris: Jules Renouard Libraire, 1862).

Duchenne de Boulogne, G. B., *The Mechanism of Human Facial Expression*, ed. by R. A. Cuthbertson (Cambridge: Cambridge University Press, 1990).

Dumas, G., *Le Sourire. Psychologie et Physiologie* (Paris: Presses Universitaires de France, 1948).

Eibl-Eibesfeldt, I., *Human Ethology* (New York: Aldine de Gruyter, 1989).

Ekman, P., *Telling Lies: Clues to Deceit in the Marketplace, Politics, and Marriage* (New York: W. W. Norton, 1985).
Ekman, P., *Emotions Revealed: Recognizing Faces and Feelings to Improve Communication and Emotional Life* (New York: Times Books, A Division of Henry Holt and Company, 2003).
Ekman, P., and W. V. Friesen, 'Constants across Cultures in the Face and Emotion', *Journal of Personality and Social Psychology*, 17 (1971), 124 - 29.
Ekman, P., and W. V. Friesen, *Unmasking the Face. A Guide to Recognizing Emotions from Facial Clues* (Englewood Cliffs: Prentice-Hall, 1975).
Ekman, P., and W. V. Friesen, *Facial Action Coding System (FACS)* (Palo Alto: Consulting Psychologist Press, 1978).
Ekman, P., and W. V. Friesen, 'Felt, False, and Miserable Smiles', *Journal of Nonverbal Behavior*, 6 (1982), 238 - 52.
Ekman, P., W. V. Friesen, and J. C. Hager, *Facial Action Coding System (FACS)* (Salt Lake City: Research Nexus division of Network Information Research Corporation, 2002).
Ekman, P., R. J. Davidson, and W. V. Friesen, 'The Duchenne Smile: Emotional Expression and Brain Physiology II', *Journal of Personality and Social Psychology*, 58, 2 (1990), 342 - 53.
Ekman, P., and M. O'Sullivan, 'Who Can Catch a Liar?', *American Psychologist*, 46, 9 (1991), 913 - 20.
Emde, R. N., and K. L. Koenig, 'Neonatal Smiling and Rapid Eye Movement States', *Journal of the American Academy of Child Psychiatry*, 8 (1969), 57 - 67.
Emde, R. N., and K. L. Koenig, 'Neonatal Smiling, Frowning, and Rapid Eye Movement States. II Sleep-Cycle Study', *Journal of the American Academy of Child Psychiatry*, 8 (1969), 637 - 56.
Emde, R. N., R. D. Mc Cartney, and R. J. Harmon, 'Neonatal Smiling in REM States, IV: Premature Study', *Child Development*, 42 (1971), 1657 - 61.
Emde, R. N., and R. J. Harmon, 'Endogenous and Exogenous Smiling Systems in Early Infancy', *Journal of the American Academy of Child Psychiatry*, 11 (1972), 177 - 200.
Ezzamel, M., and K. Hoskin, 'Retheorizing Accounting, Writing and Money, with Evidence from Mesopotamia and Ancient Egypt', *Critical Perspectives in Accounting*, 13, 3 (2002), 333 - 67.
Fitch, T. W., M. D. Hauser, and N. Chomsky, 'The Evolution of the Language Faculty: Clarifications and Implications', *Cognition*, 97 (2005), 179 - 210.
Fogel, A., and E. Thelen, 'Development of Early Expressive and Communicative Action: Reinterpreting the Evidence from a Dynamic Systems Perspective', *Developmental Psychology*, 23, 6 (1987), 747 - 61.
Fogel, A., G. C. Nelson-Goens, H. C. Hsu, and A. F. Shapiro, 'Do Different Infant Smiles Reflect Different Positive Emotions?', *Social Development*, 9 (2000), 497 - 520.
Fogel, A., H. C. Hsu, A. F. Shapiro, G. C. Nelson-Goens, and C. Secrist, 'Effect of Normal and Perturbed Social Play on the Duration and Amplitude of Different Types of Infant Smiles', *Developmental Psychology*, 42, 3 (2006), 459 - 73.
Foss, B. M., ed., *Determinants of Infant Behavior*, vol. 2 (New York: John Wiley, 1963).
Foucault, M., *The Order of Things* (London: Tavistock, 1970), first published as *Les Mots et les Choses* (Paris: Gallimard, 1966).
Foucault, M., *Discipline and Punish* (London: Allen Lane, 1977).

Fox, N. A., and R. J. Davidson, 'Patterns of Brain Electrical Activity during Facial Signs of Emotion in 10 Month Old Infants', *Developmental Psychology*, 24, 2 (1988), 230 – 36.

Frank, M., P. Ekman, and W. V. Friesen, 'Behavioral Markers and Recognizability of the Smile of Enjoyment', *Journal of Personality and Social Psychology*, 64 (1993), 83 – 93.

Gardner, A. R., and B. T. Gardner, 'Evidence for Sentence Constituents in Early Utterances of Child and Chimpanzee', *Journal of Experimental Psychology*, 104 (1975), 244 – 67.

Gillispie, C. C., *Science and Polity in France. The End of the Old Regime* (Princeton: Princeton University Press, 1980).

Gonçalves, L., W. Lee, J. Espinoza, and R. Romero, 'Three- and 4-Dimensional Ultrasound in Obstetric Practice: Does it Help?', *Journal of Ultrasound in Medicine*, 24 (2005), 1599 – 24.

Goody, J., *The Logic of Writing and the Organization of Society* (Cambridge: Cambridge University Press, 1986).

Graffi, G., 'The Problem of the Origin of Language in Western Philosophy and Linguistics', *Lingue e linguaggio*, 6, 1 (2005), 5 – 26.

Hamilton, D., *Towards a Theory of Schooling* (London: Falmer Press, 1989), 75 – 78.

Hamilton, W. D., 'The Genetical Evolution of Social Behaviour', *Journal of Theoretical Biology*, 7 (1964), 1 – 52.

Hauser, M. D., D. Barner, and T. O'Donnell, 'Evolutionary Linguistics: A New Look at an Old Landscape', *Language, Learning and Development*, 3, 2 (2007), 100 – 32.

Hauser, M. D., N. Chomsky, and W. T. Fitch, 'The Faculty of Language: What Is It, Who Has It, and How Did It Evolve?', *Science*, 298 (2002), 1569 – 79.

Hayes, C., *The Ape in Our House* (New York: Harper & Row, 1951).

Hebden, J. C., 'Advances in Optical Imaging of the Newborn Infant Brain', *Psychophysiology*, 40 (2003), 501 – 10.

Hjorstijo, C. H., *Man's Face and Mimic Language* (Lund: Student-Literature, 1970).

Hoskin, K., and R. Macve, 'The Genesis of Accountability: The West Point Connections', *Accounting, Organizations and Society*, 13, 1 (1988), 37 – 73.

Hoskin, K., and R. Macve, 'Reappraising the Genesis of Managerialism: A Re-examination of the Role of Accounting at the Springfield Armory, 1815 – 45', *Accounting, Auditing and Accountability Journal*, 7, 2 (1994), 2 – 29.

Hoskin, K., and R. Macve, 'Knowing More as Knowing Less? Alternative Histories of Cost and Management Accounting in the USA and UK', *Accounting Historians Journal*, 27, 1 (2000), 91 – 149.

Hoskin, K., and R. Macve, 'The Pennsylvania Railroad, 1849, and the Invention of Modern Management', paper presented at the American Accounting Association Annual Meeting, Orlando (FL), August 2004.

Hoskin, K., and R. Macve, 'The Western Railroad Crash of 1841: the first Manageriogenic Disaster?', paper presented at the *Interdisciplinary Perspectives on Accounting* Conference, July 2009.

Huntingford, F. A., *The Study of Animal Behaviour* (London: Chapman & Hall, 1984).

Hurford, J. R., M. Studdert-Kennedy, and C. Knight, eds, *Approaches to the Evolution of Language: Social and Cognitive Bases* (Cambridge: Cambridge University Press, 1988).

Izard, C. E., *The Maximally Discriminative Facial Movement Coding System (MAX)* (Newark: University of Delaware, Instructional Resources Center, 1979).

Izard, C. E., L. M. Dougherty, and E. A. Hembree, *A System for Identifying Affect Ex-*

pressions by Holistic Judgments (AFFEX) (Newark: Instructional Resources Center, University of Delaware, 1983).

Izard, C. E., 'Basic Emotions, Relations Among Emotions, and Emotion-Cognition Relations', *Psychological Review*, 99, 3 (1992), 561 – 65.

Jackendoff, R., 'Your Theory of Language Evolution Depends on your Theory of Language' (unpublished ms.).

Jackendoff, R., *Foundations of Language: Brain, Meaning, Grammar, Evolution* (Oxford: Oxford University Press, 2002).

Jackendoff, R., and S. Pinker, 'The Nature of the Language Faculty and Its Implications for Evolution of Language (Reply to Fitch, Hauser, and Chomsky)', *Cognition*, 97 (2005), 211 – 25.

Jones, B. E., 'Paradoxical Sleep and its Chemical/Structural Substrates in the Brain', *Neuroscience*, 40 (1991), 637 – 56.

Kail, R., ed. *Advances in Child Development and Behavior*, vol. 35 (Oxford: Elsevier, 2007).

Kako, E. T., 'Elements of Syntax in the Systems of Three Language-trained Animals', *Learning & Behavior*, 27, 1 (1999), 1 – 14.

Korner, A. F., 'Neonatal Startles, Smiles, Erections, and Reflex Sucks as Related to State, Sex, and Individuality', *Child Development*, 40 (1969), 1039 – 53.

Langford, D. J., S. E. Crager, Z. Shehzad, S. B. Smith, S. G. Sotocinal, J. S. Levenstadt, M. L. Chanda, D. J. Levitin, and J. S. Mogil, 'Social modulation of pain as evidence for empathy in mice', *Science*, 312 (2006), 1967 – 70.

Lenneberg, E. H., *Biological Foundations of Language* (New York: Wiley, 1967).

Leventhal, R., 'The Emergence of Philological Discourse in the German States, 1770 – 1810', *Isis*, 77 (1986), 243 – 60.

Lewis, M., and L. A. Rosenblum, eds, *The Development of Affect* (New York: Plenum Press, 1978).

Liberman, A. M., and others, 'Perception of the Speech Code', *Psychological Review*, 74 (1967), 431 – 61.

Manstead, A. S. R., N. H. Frijda, and A. Fischer, eds, *Feelings and Emotions* (Cambridge: Cambridge University Press, 2004).

Maquet, P., J. M. Péters, J. Aerts, G. Delfiore, C. Degueldre, A. Luxen, and others, 'Functional Neuroanatomy of Human Rapid-Eye-Movement Sleep and Dreaming', *Nature*, 383, 6596 (1996), 163 – 66.

Marks, G. A., J. P. Shaffery, A. Oksenberg, S. G. Speciale, and H. P. Roffwarg, 'A Functional Role for REM Sleep in Brain Maturation', *Behavioural Brain Research*, 69 (1995), 1 – 11.

McGrew, W. C., Review of 'Primates and Philosophers. How Morality Evolved' by F. de Waal, *PaleoAnthropology* (2007), 34 – 35.

Messer-Davidow, E., and others, eds, *Knowledges: Historical and Critical Studies in Disciplinarity*, (Charlottesville: University of Virginia Press, 1993).

Messinger, D. S., A. Fogel, and K. L. Dickson, 'What's in a Smile?', *Developmental Psychology*, 35 (1999), 701 – 08.

Messinger, D. S., A. Fogel, and K. L. Dickson, 'All Smiles are Positive, but Some Smiles are More Positive than Others', *Developmental Psychology*, 37, 5 (2001), 642 – 53.

Messinger, D. S., M. Dondi, C. G. Nelson-Goens, A. Beghi, A. Fogel, and F. Simion, 'How Sleeping Neonates Smile', *Developmental Science*, 5, 1 (2002), 49 – 55.

Mirmiran, M., 'The Functions of Fetal/Neonatal Rapid Eye Movement Sleep', *Behavioural Brain Research*, 69 (1995), 13 – 22.

Myers, M. L., *The Soul of Modern Economic Man. Ideas of Self-Interest: Thomas Hobbes to Adam Smith* (Chicago – London, University of Chicago Press, 1983).

Ong, W., *Orality and Literacy* (London: Methuen, 1982).

Oster, H., *Baby FACS: Facial Action Coding System for Infants and Young Children* (Draft version revised April 2007, in press).

Piattelli-Palmarini, M., 'Ever Since Language and Learning: Afterthoughts on the Chomsky-Piaget Debate', *Cognition*, 50 (1994), 315 – 46.

Piattelli-Palmarini, M., 'Evolution, Selection and Cognition: from 'Learning' to Parameter Setting in Biology and in the Study of Language', *Cognition*, 31, 1 (1989), 1 – 44.

Pinker, S., and P. Bloom, 'Natural Language and Natural Selection', *Behavioral and Brain Sciences*, 13, 4 (1990), 707 – 26.

Pinker, S., and R. Jackendoff, 'The Faculty of Language: What's Special about It?', *Cognition*, 95 (2005), 201 – 36.

Pinker, S., *The Language Instinct* (New York: Morrow, 1994).

Prechtl, H. F. R., and D. Beintema, *The Neurological Examination of the Full Term Newborn* (London: SIMP Heinemann, 1964).

Rizzi, L., *Issues in Italian Syntax* (Dordrecht: Foris, 1982).

Rosenberg, E., and P. Ekman, *What the Face Reveals: Basic and Applied Studies on Spontaneous Expressions Using the Facial Action Coding System (FACS)*, 2nd edn (New York: Oxford University Press, 2005).

Russell, J. A., 'Is there Universal Recognition of Emotion from Facial Expression? A Review of the Cross-cultural Studies', *Psychological Bulletin*, 115, 1 (1994), 102 – 41.

Russell, J. A., and F. M. Fernández-Dols, eds, *The Psychology of Facial Expression*, (New York: Cambridge University Press, 1997), pp. 205 – 26.

Savage-Rumbaugh, E. S., *Ape Language: From Conditioned Response to Symbol* (New York: Columbia University Press, 1986).

Schmandt-Besserat, D., *Before Writing*, 2 vols (Austin: University of Texas Press, 1992).

Schmidt, K. L., Z. Ambadar, J. F. Cohn, and L. I. Reed, 'Movement Differences Between Deliberate and Spontaneous Facial Expressions: *Zigomaticus Major* Action in Smiling', *Journal of Nonverbal Behavior*, 30, 1 (2006), 37 – 52.

Segerstrale, U., *Defenders of the Truth: The Sociobiology Debate* (Oxford: Oxford University Press, 2001).

Silk, J. B., S. F. Brosnan, J., Vonk, J. Enrich, D. J. Povinelli, A. S. Richardson, S. P. Lambeth, J. Mascaro, and S. J. Schapiro, 'Chimpanzees are indifferent to the welfare of unrelated group members', *Nature*, 437 (2005), 1357 – 59.

Smith, A., *The Theory of Moral Sentiments*, ed. by D. Daiches Raphael and A. L. MacFie (Oxford: Clarendon Press, 1976, orig. pub. 1759).

Societé de Linguistique de Paris, 'Statuts approuvés par décision ministérielle du 8 mars 1866', *Bulletin de la Société de Linguistique de Paris*, I (1871), iii-v.

Sperber, D., and D. Wilson, *Relevance: Communication and Cognition* (Blackwell: Oxford University Press, 1986; 2nd edition 1995).

Spitz, R. A., *The First Year of Life: A Psychoanalytic Study of Normal and Deviant Development of Object Relations* (New York: Basic Book, 1965).

Spitz, R. A., and K. M. Wolf, 'The Smiling Response: A Contribution to the Ontogenesis of Social Relations', *Genetic Psychology Monographs*, 34 (1946), 57 - 125.

Sroufe, L. A., and E. Waters, 'The Ontogenesis of Smiling and Laughter: A Perspective on the Organization of Development in Infancy', *Psychological Review*, 83, 3 (1976), 173 - 89.

Sroufe, L. A., *Emotional Development* (New York: Cambridge University Press, 1996).

Tattersall, I., *Becoming Human: Evolution and Human Uniqueness* (New York: Harcourt Brace, 1998).

Tian, Y.-L., T. Kanade, and J. F. Cohn, 'Recognizing Action Units for Facial Expression Analysis', *IEEE Transactions on Pattern Analysis and Machine Intelligence*, 23 (2001), 97 - 115.

Vasey, P. L., 'Interventions and alliance formation between female Japanese macaques, *Macaca fuscata*, during homosexual consortships', *Animal Behaviour*, 52 (1996), 539 - 51.

Waal, F. B. M. de, *Primates and Philosophers* (Princenton: Princeton University Press, 2006).

Walusinski, O., A. Kurjak, W. Andonotopo, and G. Azumendi, 'Fetal Yawning Assessed by 3D and 4D Sonography', *The Ultrasound Review of Obstetrics & Gynecology*, 5, 3 (2005), 210 - 17.

Warneken, F., B. Hare, A. P. Melis, D. Hanus, and M. Tomasello, 'Spontaneous altruism by chimpanzees and young children', *PloS Biology*, 5 (2007), e184.

Warneken, F., and M. Tomasello, 'Altruistic Helping in Human Infants and Young Chimpanzees', *Science*, 311 (2006), 1301 - 03.

Wilson, E. O., *Sociobiology: The New Synthesis* (Cambridge: Cambridge University Press, 1975).

Wolff, P. H., 'Observations on Newborn Infants', *Psychosomatic Medicine*, 21, 2 (1959), 110 - 18.

Wolff, P. H., *The Causes, Controls and Organization of Behavior in the Neonate* (New York: International Universities Press, 1966).

Wolff, P. H., The *Development of Behavioral States and the Expression of Emotions in Early Infancy: New Proposals for Investigation* (Chicago: University of Chicago Press, 1987).

Wray, A., ed., *The Transition to Language* (Oxford: Oxford University Press, 2002).

Part III: Quests for Origins, Origins of Narrations

Abstracts

Marcella Farioli

The Genesis of the Cosmos, the Search for the *Arche* and the Finding of *Aitia* in Classical Greek Culture

Every archaic civilisation addressed the problem of the origins of the gods, the cosmos and man. Greek enquiries are characterised by a marked aetiological tendency, which firstly emerged in mythology and folk-tales, then in philosophy, historiography, medicine, and the natural science, and eventually, during the Hellenistic Age, in the writings of poets and philologists.

Greek representations of origins fit within four typologies. The first includes cosmogonies, that is mythological stories about the origins of the world – Hesiod's *Theogony*, Orphic cosmogonies, genealogical manuals – based on older patterns deriving from the Ancient Near East. The second typology comprises speculations about the *arche*, the first cause of the world, by Ionic philosophers. The substantial element of novelty they introduced was the idea that it is necessary to go beyond phenomena to detect, through observation of nature and abstract thought, the unifying principle of the cosmos. The third kind of representation focuses on the birth of the human race and the origin of evil, two issues very much present in Greek mythology, i. e. in Orphic thought and in the myths of Prometheus and Pandora. Surprisingly, the Greeks did not have a specific founding myth for the creation of man. Such an absence is likely due to the fact that for them the main characteristic of mankind was not its existence, but rather its mortality. The lack of a unified anthropogony is balanced by a widely shared anthropology, which explains when and why men have become what they are. Finally, the issue of origins is expressed in several aethiological myths relating to *protoi heuretai* (inventors) and the *aitia* (origins) of objects, phenomena, and institutions. The *Aitia* became an autonomous literary genre in the Hellenistic Age.

Alessandro Zironi

Searching for Origins: Indo-European Words and Nordic Representations

Indo-European words which indicate 'origins' evoke stasis or kinesis, call attention to the starting point or to the movement they engender, focus on the beginning or on the outcome. The double interpretation of the concept of origin can be retraced in the Nordic myth of the creation of cosmos, which narrates how the primeval void called *ginnungagap* was filled.

Old Norse cosmogony defines a model of origins which has constantly been associated with representations of Iceland. The unceasing outburst of primary elements has left deep marks and scars on Icelandic landscapes, and has prevented human beings from prevailing over, and moulding, nature's primary manifestations. The interest for Iceland has thriven on the observation of caves, abysses and volcanoes, thus fostering speculations on what can be defined as primeval.

The connection between the Icelandic landscape and the search for origins is further examined in relation to *Voyage au centre de la Terre* (1864) by Jules Verne and to contemporary artistic representations.

Marusca Francini

Origo Gentis and Cultural Memory. The Origins of England as an Emblematic Case Study

The focus of this chapter is on the origins of Anglo-Saxon England and the narrations they elicited in the two major historical works of the Anglo-Saxon period, namely the *Historia Ecclesiastica Gentis Anglorum*, written in Latin by the monk Bede in the eighth century, and the *Anglo-Saxon Chronicle*, a set of annals written in Old English in the ninth century.

Two different approaches to the foundation myth of Anglo-Saxon England – the Anglo-Saxon invasion of Britain in the fifth century – will be compared in order to demonstrate that for Bede the idea of a united England was based on his belief in the Anglo-Saxon Church, whereas patriotism in the *Anglo-Saxon Chronicle*, which is a secular work, was rooted in the idea of victorious military kingship.

Marianne Børch

The Semantics of Originality as Indicator of the Cultural Paradigm Shift from Medieval to Modern

Today, the word 'original' can be regarded as a synonym for 'innovative' and 'individual'. This usage, however, is directly inverse to that of the Middle Ages and the Renaissance, when 'original' meant 'pristine' and 'ideal' (as in the first and total 'Original Sin'). Denoting the absolute, 'original' referred to the standard against which everything else was measured, every departure from that standard being judged by various degrees of 'lessness' or degeneration.

A focus on the works of Geoffrey Chaucer will allow for demonstrating how the apparent simplicity, occasional strangeness and referential inconsistency of early literary texts and other art forms may be explained in terms of their closeness to an absolute, original norm. For example, the cliché of romance knighthood, which seems to describe all knights as alike, reflects the attempt to depict an ideal society, where the individual coincides with the ideal. Moreover, narrative inconsistencies in characterisation (a kind husband who is nevertheless a 'demon') may be related to the difference between the character's function with regards to medieval faith in God's indulgence (which requires the 'ideal' role of a pagan) and his function in a particular story (which requires a loving husband).

The chapter finally explores how literary texts testify to the epistemological change from medieval to modern through words which originally functioned as ideal referents and later acquired socially determined meanings.

Jón Gunnar Þorsteinsson

Paradigm as an Origin: Or, Bearing Evidence to the Meaningful Logic of Cause and Effect

This chapter questions the relevance of origins as a paradigm by developing a comparison between the concept of Author in French Post-Structuralism and recent cosmological theories. The origin of a river, of cartoons and of Modernism will also be discussed in order to show that origins can be an elusive paradigm and the quest for origins a deceitful enterprise.

Searching for origins of phenomena and concepts can move us away from them, because it is usually possible to go one step further back, and from that step, another one backwards. Such process can lead any serious analysis astray.

The assumption that there is a precise beginning which is more meaningful than that which follows can undermine the chain of cause and effect. Although we can often trace the origin of phenomena much further back, the process can become pointless. Once the concept of origins is enlarged too much, meaningful relations are breached. At risk is that we get lost in arguing about definitions which are perhaps not relevant in themselves, instead of doing work on the actual phenomena. In order to avoid the risk of undermining the cause-and-effect nexus, it is important to assess when going further backwards would be irrelevant.

Marcella Farioli

The Genesis of the Cosmos, the Search for the *Arche* and the Finding of *Aitia* in Classical Greek Culture

In every archaic civilization people questioned how things had come to be as they were. The issues surrounding the origins of the divine order, of the cosmos and of man constitute the main lines of inquiry. Indeed, in every culture there are mythical stories about the origins of the universe and it is into these that different peoples and societies have introduced the elements and metaphors most representative of their world view.

Starting from 3000 BCE the first cosmogonies, theogonies and anthropogonies, which narrate the origins of the universe and are based essentially on the authority of myth, were codified and spread throughout Egypt and Mesopotamia. In Greece different theogonies and cosmogonies dating to the eighth century point to an earlier mythological inheritance, also of Eastern origin. Later, for several historical, cultural and economic reasons, the search for the *arche*, the origin and the first principle of all things, developed in Ionia, connecting the traditional tale and later philosophical conceptions.

Although the issue of origins is typical of all archaic cultures, among the Greeks it was particularly pressing and endured for centuries, even after philosophy would tackle it on rational ground. Overwhelmed by the fear of *horror vacui*, they needed to explain and define the genesis of the greatest possible number of deeds, objects and phenomena, to show their *aition*, their origins, to place them in space and in time and find a discoverer or inventor, the *protos heuretes*. The quest for origins allowed the Greeks to express a deep connection with their own roots, with the cult of memory and the excitement of the founding act.

For these reasons the inquiry into the origins of a noteworthy number of objects, phenomena, events and rites establishes and transforms itself throughout the development of Greek civilization: such marked aetiological tendency [1] is to emerge in historiography, in medicine, natural science and, eventually, in the Hellenistic Age, in the scholarship of poets and philologists.

1 See Bernhard Abraham van Groningen, *In the Grip of the Past. Essay on an Aspect of Greek*

I. Greek and Ancient Near East Cosmogonies

In archaic Greek thought the idea of cosmogony as conceived by modern science, that is to say a scientific study about the origin and evolution of the universe, did not exist. There were, instead, cosmogonies dressed up as mythological stories about the origins of the world.

The first cosmogonic treatment that has reached us in its entirety is the *Theogony* by Hesiod, a poem in hexameters written at the end of the eighth century BCE. It is also one of the few sources in our possession which expresses Greek cosmogonic thought prior to Milesian conceptualizations of *arche*. A long list of divine filiations begins with Chasm and continues with Gaia, the earth, Tartarus, the world's deepest region, and Eros, that is Love. Chasm himself generates two dark entities, Erebus, infernal darkness, and Night, whose union generates two opposing luminous entities, Aether, the bright, lit part of the sky, and Day. Even Gaia, through parthenogenesis, gives birth to a son, Ouranos, the sky, with whom she also produces Oceanus:

> In truth, first of all Chasm came to be, and then broad-breasted Earth, the ever immovable seat of all the immortals who possess snowy Olympus' peak and murky Tartarus in the depths of the broad-pathed earth, and Eros, who is the most beautiful among the immortals gods [...].
> From Chasm, Erebos and black Night came to be; and then Aether and Day came forth from Night, who conceived and bore them after mingling in love with Erebos. Earth first of all bore starry Sky, equal to herself, to cover her on every side, so that she would be the ever immovable seat for the blessed gods; and she bore the high mountains, the graceful haunts of the goddesses, Nymphs who dwell on the wooded mountains. And she also bore the barren sea seething with its swell, Pontus, without delightful love; and then having bedded with Sky, she bore deep-eddying Ocean [...]. [2]

The story of Ouranos' castration, which is performed by his own son, Cronos, is then narrated and followed by further filiations, among which the birth of the Olympian Gods, whose parents are Cronos and Rhea, and the fight between Zeus and his father Cronos, who is eventually dethroned. Zeus' accession to the throne is seen by the poet as the moment of change from violence and disorder to order and justice. Afterwards, the genealogy of the Gods lists the minor gods, the

Thought (Leiden: E. J. Brill, 1953), pp. 30–31.

2 See Hesiod, *Theogony, Works and Days, Testimonia*, ed. and trans. by Glenn W. Most (Cambridge – London: Harvard University Press, 2006), ll. 116–20; 123–33, p. 13. In note 7, ivi, Most explains that Chasm is 'usually translated 'Chaos'; but that suggests to us, misleadingly, a jumble of disordered matter, whereas Hesiod's term indicates instead a gap or opening'.

unions between the Gods and the unions between the Gods and the mortals, which will give rise to the heroes.

The gods described by Hesiod before the appearance of the Olympians have a complex physiognomy: they are abstract, cosmological entities (the Earth, the Sky) who nevertheless behave anthropomorphically, coupling and procreating. How abstract elements and anthropomorphic behaviours intermingle is not specified: Hesiod's text, in fact, does not supply precise answers to the question of origins or to how such births happened. How was the earth formed? First there was Chasm and then there was the Earth. And Chasm, what is it, and where does it come from? [3]

The *Theogony* is certainly a story and not a scientific explanation. And like most archaic mythical cosmogonies, it consists of a juxtaposition of events ('in the beginning there was [...] then something became or happened'), while others are based on a technique of subtraction ('when the sky was not yet and the earth did not exist').

The choice of the narrative pattern did not derive from a deliberate disinterest in 'scientific' subjects. Firstly, Hesiod's intention testifies to the ethical and religious need to describe the order established by Zeus; secondly, and even more simply, for Hesiod and his audience the narrative form of the mythological story was a natural way to express and understand the Universe. Listeners perceived the elements of the story as true, but they did not understand exactly what happened from a point of view that we would define as 'scientific' or 'rational'. Hesiod's perspective was not a pre-philosophical cosmogony, as some have defined it, because the poet's focus was not on the physical nature of the things in the world, but rather on their origins in connection with the gods.

A distinctive feature of the ancient mythical cosmogonies and, though with differences, of the Ionic philosophy which followed was an undefined first principle which exists at the origin of the Universe and from which successive determinations are derived: this 'concept of principle, a principle from which everything must spring, must be seen as a speculative conquest'. [4] It is Chasm for Hesiod, Night in the Orphic theogony, but it is often identified with the primeval waters, as in many Ancient Near East's myths and in Homer. The Homerical poems do not show particular cosmogonic interests. Only in two passages of the *Iliad* do these sort of topics emerge; they are related especially to water. In the first of the two groups of verses Hera states:

3 How Hesiod imagined Chasm we do not know exactly (a void? the place in which things came into being?). See Hesiod, *Theogony*, ed. with Prolegomena and Commentary by Martin L. West (Oxford: Clarendon Press, 1966), pp. 192–93.

4 Walter Burkert, *Da Omero ai Magi. La tradizione orientale nella cultura greca* (Venezia: Marsilio, 1999), pp. 47–48.

For I am faring to visit the limits of the all-nurturing earth, and Oceanus, from whom the gods are sprung, and mother Tethys [...]. [5]

Oceanus, the river god that surrounds the world, and Tethys, the female goddess of the waters (and who should not be confused with the sea goddess Thetis) would be the first couple at the beginning who begat all others. A few verses later Sleep speaks of the streams of river Oceanus, from whom they are all sprung. [6] A nearly unknown goddess in Greek mythology, Tethys' mythical physiognomy, whatever it was, is revealed by comparative analysis. A long tradition, based on the cultural bias of Hellenic superiority over other peoples, has fuelled the belief in the originality of Greek cosmogonies; on the contrary, as shall be seen in relation to pre-Socratic philosophy, they are in large part modelled on much older prototypes which came from the Ancient Near East.

Water as a cosmogonic principle is well documented East of Greece: in the Babylonian poem *Enûma elish*, for instance, the paradigmatic text about the origin of the world, Apsu, the fresh water ocean, and Tiamat, the salty water sea, represented the father and the mother of all things at a time when the sky and the earth did not exist, yet. Just like Oceanus and Tethys in the *Iliad*: indeed, as many scholars have noted, [7] the quoted Homerical passage testifies to the influence of Babylonian mythology on archaic epic.

The topic of primeval waters is present in different forms in Vedic literature and in Egyptian civilization. The latter began formulating a well-structured cosmogony around 3000 BCE. This had a non-uniform development and was tied to different theocratic centres. Many of these cosmogonies set water as the cosmogonic principle: Nun, the primeval abyss, the great murky lake which contained everything, interrupted its indistinct sleep at a certain point. In the Heliopolitan cosmogony, for example, Nun gives birth to the sun god, Atum, before the birth of the sky and of the earth takes place. Having climbed to the top of a hill, with his spit he creates Shu, the Void, and the goddess Tefnut, Humidity, who in turn together beget Geb and Nut, the earth and sky. From these two, two pairs of siblings were born, Osiris, Isis, Seth and Nefti, all of whom together give birth to humanity. In the Hermopolitan theology, instead, a hill of mud emerges from the water, generating eight primeval gods, four masculine with a frog head

5 Homer, *The Iliad*, 2 vols, ed. by Augustus T. Murray (London, W. Heinemann – Cambridge, Mass., Harvard University Press, 1934–37), II, Book XIV, ll. 200–01, p. 81.

6 Ibid., XIV, ll. 245–46, p. 85.

7 See Burkert, pp. 14–30. Analogously, in the Accadic poem *Atrahasis* (seventeenth century BCE) we find at the origin of the world the division of the cosmos in three parts attributed to three gods, just like in *Iliad*, XV, ll. 187–93, in which the three kingdoms are given to Zeus, Ades and Poseidon.

and four feminine with a snake head who would later produce the successive generations.

The unity of an undefined element at the origin is followed by separation: the first one is often light - darkness as in *Genesis* 1, in which the unformed Earth is imagined in a sort of *continuum* with murky darkness, from which the earth - sky is separated through the first act of creation. Like the Hittite myth of Kumarbi who swallows the phallus of the sky god, Anu, the Greek myth of the castration of Ouranos, the personification of the sky, practised by Zeus and instigated by Gaia, the earth, has thus the function of explaining the definitive separation of the original couple sky - earth.

After the initial separation, two possible developments, or two storytelling genres, exist in the mythical stories; they can be defined as biomorphic and technomorphic. [8]

The biomorphic pattern introduces couples of different genders who reproduce, giving life to successive generations which clash with each other, such as in the fight between Ouranos and Cronos and between Cronos and Zeus in Hesiod. Similar 'succession myths' can be found in the *Enûma Elish* and in the myth of Kumarbi.

The technomorphic pattern focuses on a demiurge who shapes matter and gives origin to things, as in *Genesis* or in many of the demiurge god figures in *Rigveda*. Nevertheless, 'the idea of a created world is untypical of early Greek thinking': [9] at this time the idea of a creation *ex nihilo* is absent; the figure of a Supreme Creator and the problem of what existed prior to creation are also absent. The pure technomorphic pattern makes itself known only with the Platonic demiurge of *Timeus*, who actually does not create *ex nihilo*, but shapes and orders pre-existent matter. [10] Before that we find, at best, a co-existence of the two storytelling frames, in various combinations, in the thought of some pre-Socratic philosophers, or, for example, in the Orphic cosmogony of Derveni's papyrus.

Dating to the fourth century, this papyrus contains a late fifth century text which itself includes the exegesis of a presumably older Orphic poem. The cosmogony of Derveni's papyrus begins with Night, followed by Ouranos as the first king, and reconnects with Hesiod by referring to the episode of Ouranos's castration. Zeus is here celebrated as the first and the last, the principle, means, aim, lord of everything: the god who creates first the gods, then the world, by

8 See Burkert, pp. 49 – 52.

9 Martin L. West, *Indo-European Poetry and Myth* (Oxford: Oxford University Press, 2007), p. 354.

10 The same also happens in *Genesis* 1, later interpreted as creation *ex nihilo* in the light of *Genesis* 2. Furthermore, in contemporary primitive societies we find these two models and a third creationist model based on the figure of a supreme creator.

accomplishing sexual acts, as in traditional theogonies, and 'by thinking' (*mésato*). We have, therefore, a co-existence of theogony-cosmogony and creation through deliberate mental acts: this co-existence of the two models of the universe's origins is also typical of other Indo-European cultures and is found, for example, in the *Enûma Elish.* [11]

Another source for the Orphic cosmogonies is a famous passage in Aristophanes' *Birds*, where the choir narrates its own origins in the so-called 'ornithogony':

In the beginning were Chaos and Night and black Erebus and broad Tartarus, and no Earth, Air or Sky. And in the boundless bosom of Erebus did black-winged Night at the very start bring forth a wind egg, from which as the seasons revolved came forth Eros the seductive, like to swift whirlwinds, his back aglitter with wings of gold. And mating by night with winged Chaos in broad Tartarus, he hatched our own race and first brought it up to daylight. There was no race of immortal gods before Eros commingled everything; then as this commingled with that, Sky came to be, and Ocean and Earth, and the whole imperishable race of blessed gods. [12]

Since this is a comical text, it is not possible to use the passage *in toto* as a source to retrace Orphic cosmogonic thought; nevertheless, it seems obvious that it represents a mixture of comical invention and genuine, ancient and contemporary, cosmogonic traditions wherein Orphic and Hesiodic influences can be noted.

The idea of a cosmic egg from which things took their beginning is not only found in Orphism. This mytheme is actually documented in many archaic cultures, and not only: it is present in the mythology of many populations in Africa, China, India and Japan. In its cosmogenic content, it symbolizes the primeval unity of being, the perfect, undivided whole which comes before the separation of the elements and the birth of the perceptible Universe.

We also find this mytheme in the Hindu culture. The cosmogonic stories of Hinduism are many and appear in the beginning in the *Vedas*, then in the *Brahmana*, the *Upanishad*, the epics (*Mahabharata* and *Ramayana*) and in the *Puranas*; some of these are real myths, whilst others have a more speculative and philosophical character. In great part, however, Brahma appears as a demiurge: Hindu myth did not consider creation *ex nihilo*, but rather provided a framework and organization of the elements which made up the universe.

The most widespread myth says that at its beginning the cosmic egg, Hir-

11 Burkert, pp. 78 – 86. About the co-existence of the two models see also West, *Indo-European Poetry*, pp. 354 – 59.

12 Aristophanes, *Birds*, in *Birds, Lysistrata, Woman at the Thesmophoria*, ed. by Jeffrey Henderson (Cambridge – London: Harvard University Press, 2000), ll. 693 – 703, p. 117.

anyagarbha, floated in the primeval ocean wrapped in the darkness of non-existence. When the egg hatches, from the upper half of the shell, which is made of gold, the sky is born; from the lower half, which is made of silver, the earth is born. The internal membranes of the shell form the mountains and the external ones form the clouds; the veins and the liquids give life to the rivers and seas.

A similar tale also appears in archaic Chinese myth: it is the story of Pan Gu. At the beginning of the world, the sky and the earth formed a great egg, in which Pan Gu grew up and became a giant. He broke the shell in two parts with an axe, thereby dividing the sky from the earth. Pan Gu grew even more and every part of his body transformed itself: chest, stomach, arms and legs changed into mountains, the eyes became the Sun and the Moon, the clouds and wind were made from his breath. His fleas became men and the other four types of living beings: animals, plants, spirits and ghosts. It is evident that also in this myth the egg represents the original unity of the world, where the Sky and the Earth are still undivided.

The idea of the body which divides to give life to the elements of the cosmos is also present in Babylonian myth, wherein the goddess Tiamat is split in two either as a fish to dry or as an oyster or an egg, and from her body parts the Sky, the Earth and the other elements of the universe are made. Seen in a more general way, the idea that Creation is the result of a violent crisis, a cruel sacrifice or death is present in several cosmogonic myths, not only in the Indo-European area, as testified by the slaying of the primeval monster Vrtra, the dismemberment of Prajapati and of the giant Purusha in the *Rigveda*, and the killing and dismemberment of the giant Ymir in Nordic mythology.

The last known category of Greek cosmogonic texts from the archaic period includes genealogical manuals in prose and poetry, of which we possess only fragments. Information from later sources is also relevant, even though not always reliable, because their Neo-Platonic inspiration sometimes drove the authors to interpret the ancient texts according to later philosophical categories.

Of these authors, who can be placed between the seventh and the sixth century BCE, we know little more than the name of one, Acusilaus, who wrote prose genealogies in which Chaos was presented as the first principle. Museus also composed a *Theogony* in which the origin of things was traced back to Tartarus and Night, whilst Epimenides, author of another *Theogony*, identified it as Air and Night.

Better known is the thought of Pherecydes of Syrus (sixth century) who, it seems, created an original cosmogony from the union of Greek and Oriental elements: it started from three original gods, Zas (Zeus), Chronos (Time), and Ctonia (Earth). Zas and Ctonia merge, while Chronos reproduces without a female counterpart, creating from his own semen Fire, Air and Water. The figure of Time, Chronos (not Cronus) is without precedent in the Greek cosmogonies, but is present as cosmic progenitor in the East. [13] In Zoroastrian cosmogony, for example, Zurvan, that is Time regarded as a primordial entity, reproduces by having sexual intercourse with himself, just like Kâla, the Indian god of Time, does. In all three cases, Time has always existed; it is not the architect of the material world, but the ancestor of a demiurge god, namely Zas in the case of Pherecydes.

According to Aristotle (*Metaphysics*, 1091b), Pherecydes' cosmogony is characterised by a mixture of mythological language and a need to rationalize; however, such combination was not successful in later traditions, perhaps because it presents mythical stories without a real 'rational' basis just at the moment when a different need to explain the world in a naturalistic manner was being established in Ionia.

II. The Search for the *Arche* in Ionic Philosophy

The origin of Greek philosophy is traditionally placed on the Ionic shores at the beginning of the sixth century: here some thinkers, Thales, Anaximander and Anaximenes would give life to speculations in materialistic, and no longer mythological, terms about the origins of the world, its founding principle and first motive, the *arche*. Aristotle himself, while examining his predecessors in Book I of *Metaphysics*, defines these thinkers as 'the first philosophers':

> Of the first philosophers, then, most thought the principles which were of the nature of matter were the only principles of all things. That of which all things that are consist, the first from which they come to be, the last into which they are resolved (the substance remaining, but changing in its modifications), this they say is the element and this the principle of things [...]. [14]

The need for such a separation from myth, even though less radical than was thought for a long time, can be attributed to several, complex and widely discussed reasons, some of which are of an economic-productive nature. In the sixth century on the Ionic shores new economic and trade models became established and, owing to the practical needs of the emerging classes, knowledge

13 Martin L. West, *Early Greek Philosophy and the Orient* (Oxford: Clarendon Press, 1971), pp. 28–75.

14 Aristotle, *Metaphysics*, trans. by William David Ross, Book I, Part 3, 983b, http://www.classicallibrary.org/aristotle/metaphysics/index.htm [accessed 9 January 2010]. See also *The Works of Aristotle. Metaphysica*, under editorship of John Alexander Smith and William David Ross (Oxford: Clarendon Press, 1908).

began to become democratic. The merchant acquired skills through observation of reality, the artisan modified nature through new materials and techniques.

Moreover in Greece, unlike in the Ancient Near East, the institutions which ought to defend myth as the foundation of authority did not exist: the priest caste, the temples and the 'house of tablets', whose knowledge rested on respect for tradition and, therefore, represented a form of power, were absent. Because these structures were missing in Greece, knowledge was transmitted in different ways, allowing for more intellectual freedom and a progressive separation from tradition. Nonetheless, the contribution of Eastern thought to Ionic philosophy was more conspicuous than scholars thought when classicist idealization of 'Hellenic genius' and its purity predominated.

The first Ionic philosopher, Thales, maintained that the generating principle, the *arche* of all things, was water; it spread throughout nature and living beings. For this reason, Thales also stated that the earth floats on water, 'getting the notion perhaps from seeing that the nutriment of all things is moist, and that the heat itself is generated from the moist and kept alive by it (and that from which they come to be is a principle of all things)'. [15]

A view, after all, not so different from previous mythical cosmogonies, so much so that Aristotle himself, a little further on, compares Thales' conception to that of 'the ancients who lived long before the present generation and first framed accounts of the gods', considering Oceanus and Tethys the 'parents of creation'. [16] Thales looks for the supreme unity of the cosmos not in a divine power, like myth does, but in a living natural principle; this difference is not to be overestimated if we consider that in that period the separation between the natural and the supernatural world was not yet perceived very clearly. It is remarkable, however, that in order to explain the multiplicity of reality, Thales avoids anthropomorphic and genealogical representations by enunciating a single principle bound to an experimental reality.

As for the second philosopher of the Miletus school, Anaximander, the *arche* from which all particular beings come is not water, but something more indefinite, the *apeiron*, that is to say the infinite and timeless composite unity of all that exists. From its perpetual motion everything arises, marking a separation from the original indistinction. Everything pays a price for such a separation, because its individual existence entails a temporal finitude and an opposition with other individual existences. The idea of the balance of opposites introduced by Anaximander and drawn from the observation of natural phenomena (air, water, fire, earth) was to strongly influence later naturalistic thought.

Anaximenes also tries to explain phenomena by providing a general outline,

15 Ibid., ivi.
16 Ibid., ivi.

thus softening the divine component. While exploring the first unity of the cosmos, he speculates on the double meaning, both meteorological (*aér*, air) and physiological (*pnéuma*, breathing, breath of life), of air. Air seems to be the only principle able to unite heavenly and biological phenomena in a single system and to connect life with the general order of the cosmos. Air extends without any limit around the world, it is eternal and in perpetual motion. All distinct beings originate from the two eternal processes of rarefaction and condensation.

The conceptions of Thales, Anaximander and Anaximenes do not form a school as we understand it and do not derive directly from each other. Moreover, as has often been noted, their views are not thoroughly original: influences range from the Greek cosmogonic tradition and the wisdom literature, including the idea that the cosmic order is a form of justice, to Eastern speculation and myths, which reached Ionia through Persia. [17]

It is not our aim to discuss the age-old issue regarding Ionic philosophical thought, which involves assessing the degree of originality and novelty in explanations of phenomena based on the material nature of the first principle, in a systematic use of 'rationalistic' knowledge, in explanations of origins in non-anthropomorphic terms, in the emphasis on experience, or in the secularization of the creation myths. It is certain that speculations about the origins carried on by these philosophers contain a substantial element of novelty: it is the idea that we have to go beyond phenomena in order to grasp, through abstraction, a principle and a unifying framework for the cosmos. The observation of *physis*, nature, allows us to reconcile the unity of the universe with the seemingly irreducible multiplicity of human experience. As West aptly observes, assumptions about the existence of universal processes are the distinctive marks of every scientific approach. [18]

Though archaic and still bound to mythical cosmogonies – even if some scholars see within them the germs of specific modern scientific theories –, the Milesian doctrines attempt to offer a rational explanation for all beings and their origins through a unifying principle of the world. Greek and Eastern mythical elements, although present, are used for the first time to answer the question: what is the primary motive of reality? Moreover, ancient cosmogonies, while offering thorough clarifications about the origins and functioning of the world, did not enquire whether the cosmos they gave an account for was 'all that exists', and consequently left its time and space limits undefined. The Milesians ad-

17 On the subject see Burkert, pp. 35 – 37. For an examination of Eastern influence on Milesian thought see *Early Greek Philosophy*. The sixth century is actually the time of the greatest influence of Eastern thought in Ionia.

18 *Early Greek Philosophy*, pp. 80 – 81.

dressed this question, postulating the infinite spatial and temporal nature of the universe, seen as a whole.

The famous passage from *mythos* to *logos* attributed to the Ionic philosophers does not represent so much the dismissal of mythical categories in order to achieve a form of logical and rational thought as it is intended in the modern age, which would be anachronistic, but rather highlights the use of experience to later reach abstract thought while simultaneously departing from the unavoidable models of the origins created by Homer and Hesiod. Eventually, in the differences between the diverse theories and in the debate they fuelled we can detect early signs of philosophy and modern science.

The above-mentioned search for the causes and principles of reality through a methodical approach based on experience was carried out during the sixth and fifth centuries BCE not only in the fields of natural science such as medicine and biology, but also in the area of human science, above all in ethnography and historiography.[19] In medicine – in the second half of the sixth century the Knidos school was born, and not by coincidence, in Asia Minor – empirical observation is established, although in an embryonic way, as the foundation for searching the origin of illness. This method was then improved in the fifth century by Hippocrates and the school of Kos. Similarly, Herodotus detached from the concept of history as past memory, developing the idea, peculiar to naturalistic and medical inquiry (*historíe*), that the origin and the aetiology of the present can be discovered in the past. Historiography thus offers hermeneutic tools. Herodotus was not able to achieve his aetiological aims, but the very fact that he posed the question was a novelty. Thucydides' historical method, instead, is anamnestic and aetiological and as is well known has several points in common with that of Hippocratic science.

III. The Birth of the Human Species and the Origin of Evil

Hesiod's *Theogony*, well detailed on the gods' genealogy, does not give any account for the birth of the human species: in the *Works and Days* the poet only tells us that 'the gods and mortal human beings came about from the same origin'.[20] In the following lines he adds that 'golden was the race of speech-endowed human beings which the immortals, who have their mansions on Olympus, made first of all. They lived at the time of Cronus, when he was king in

19 For a joint examination of the development of these two scientific branches see Mario Vegetti, 'Le scienze della natura e dell'uomo nel V secolo', in Ludovico Geymonat, *Storia del pensiero filosofico e scientifico*, 9 vols (Milano: Garzanti, 1975), I, pp. 110–50.

20 Hesiod, *Theogony, Works and Days, Testimonia*, l. 108, p. 95.

the sky'.[21] Here starts the famous 'myth of the races', which goes on to describe the five lines of mankind: the golden one, characterized by the automatism of the fruits of the Earth and by the lack of illness and old age; the silver one, which Zeus destroyed for its arrogance; the bronze one, born from its ashes, strong and war-like; the line of the heroes who shared divine ancestry: they perished in the wars and were placed on the Islands of the Blessed Ones; and eventually the descendants of the iron line, contemporaries of the poet, enduring labour and worries.

Whether the five ages foreshadow a decay, or what the chronological framework and the influences may be, are issues related to the many interpretations which this myth has fuelled: here it is worthwhile pointing out that 'how' men came to the world is a question that the Hesodic passage left unsolved.

Some traditions bind the origin of men to the Titans, Ouranos and Gaia's children, without giving, in this case as well, any detail about their genesis.[22] Furthermore, different regional variations exist in Greece and in Asia Minor, often reported by later sources, of stories about the creation or the spontaneous birth of man. They usually depict a primeval man born directly from the Earth, which, therefore, turns out to be the mother of all mortals. Here is how Hippolytus of Rome (second-third centuries BCE) shows such a multiplicity of stories:

> Now earth, say the Greeks, gave forth a man, (earth) first bearing a goodly gift, wishing to become mother not of plants devoid of sense, nor beasts without reason, but of a gentle and highly favoured creature. It, however, is difficult, (the Naassene) says, to ascertain whether Alalcomeneus, first of men, rose upon the Boeotians over Lake Cephisus; or whether it were the Idaean Curetes, a divine race; or the Phrygian Corybantes, whom first the sun beheld springing up after the manner of the growth of trees; or whether Arcadia brought forth Pelasgus, of greater antiquity than the moon; or Eleusis (produced) Diaulus, an inhabitant of Raria.[23]

Also Plato attributes a similar story to Aspasia, meant to give to the Athenians the attribute of *gegeneis*, 'born from the Earth':

---

21 Ibid., ll. 109–12, p. 97.

22 Cf. *Homeric Hymns*, trans. by Jules Cashford with an Introduction and Notes by Nicholas Richardson (London – New York: Penguin, 2003), ll. 334–36, p. 45: 'Hear me now, Gaia, and broad Ouranos above, and you Titan gods who live beneath the earth around great Tartarus from whom gods and men come'.

23 For the translation of *Refutatio omnium haeresium* by Hippolytus of Rome see *Ante-Nicene Fathers*, 10 vols, ed. by Alexander Roberts, James Donaldson, and Arthur Cleveland Coxe (Buffalo, NY: Christian Literature Publishing Co., 1886), V. *The Fathers of the Third Century*, Hyppolitus, *Refutation of All Heresies*, Book V, Chapter 2, p. 48.

During that period in which the whole earth was putting forth and producing animals of every kind, wild and tame, our country showed herself barren and void of wild animals, but chose for herself and gave birth to man, who surpasses all other animals in intelligence and alone of animals regards justice and the gods. [24]

In some such stories ash trees or ash tree Nymphs also appear, as in Phoroneus' case, the Argive first man, whilst the Thebans' ancestors, the Spartos, were also thought to be *gegeneis*, to have sprung completely armed from the earth, ready to fight each other, like Bronze Age men.

Only more recent stories make reference to a real creation of man. Plato says in *Protagoras*:

There was once a time when there were gods, but no mortals creatures. And when to these also came their destined time to be created, the gods moulded their forms within the earth, of a mixture made of earth and fire and all substances that are compounded with fire and earth. When they were about to bring these creatures to light, they charged Prometheus and Epimetheus to deal to each the equipment of his proper faculty. [25]

Nevertheless, Epimetheus gives out all faculties to the animals and man is left naked and harmless. Therefore Prometheus steals fire and technical knowledge from the gods and gives them to man, thus supplying him with a means of support. Other sources say that Prometheus moulded men with water and earth. [26]

The story about the theft of fire is also told in Hesiod (*Theogony*, ll. 535 – 616, and *Works and Days*, ll. 42 – 82). After Mecon's deceit, when Prometheus, dividing a bull between men and gods, gave the bones and fat to Zeus and the meat to the men, the latter were then deprived of fire: Prometheus thus stole it and gave it to mortals. Zeus, enraged, sent mankind a great evil, Pandora, the first woman, moulded from earth thanks to Hephaestus. As is well known, Pandora was then to open the forbidden box given by Zeus, releasing onto the earth labours, illnesses and death. By this way the stealing of fire and the origin of

24 Plato, *Menexenus*, in *Timaeus, Critias, Cleitophon, Menexenus, Epistles*, ed. by Robert G. Bury (Cambridge – London: Harvard University Press, 1929), 237d, p. 343.

25 Plato, *Protagoras*, in *Laches, Protagoras, Meno, Eutydemus*, ed. by Walter R. M. Lamb (Cambridge – London: Harvard University Press, 1924), 320 c-d, p. 129.

26 See Apollodorus, *The Library*, with an English translation by James Frazer, 2 vols (Cambridge, Mass.: Harvard University Press, 1976 – 1979), I, 7, 1. Pausanias mentions great blocks of stones in Phocis which, it was told, exhaled human smell. They were the remnant mud by which Prometheus had made the man. See Pausanias Periegeta, *Description of Greece*, with an English translation by William H. S. Jones, 5 vols (Cambridge, Mass.: Harvard University Press, 1965 – 1978), X, 4, 4.

women coincide with the origin of evil, the motive of which is another major and universal inquiry of mankind.

The relationship between the origin of man and of evil is also present in the Orphic anthropogony and is linked to an act of *hybris* of the Titans:[27] after killing Zeus and Persephones' son, little Dionysos Zagreus, they devoured his flesh. Zeus strikes them down and from their ashes men are born. Thus they take part in Dionysos' divine essence as well as in brutal titanic violence.

At a first glance it is amazing to discover that the Greeks did not have a specific founding myth for man's creation. Such absence is likely due to the fact that for the Greeks the main characteristic of men is not their existence, but rather their mortality. Consequently, the lack of a structured, shared anthropogony is balanced by a widely shared anthropology which, by explaining when and why men have become what they are, is strictly connected with the myth of Prometheus. Not only does it explain the origins of sacrifice, cooking food, marriage, labour and the necessity of evil in the world, but it also offers reasons for the definitive split between men and gods and between men and beasts: the occurrence of such a split accounts for man's creation.

IV. *Protos Heuretes* and *Aitia*

In the way described above, the myth of Prometheus is an aetiological account and the titan a *protos heuretes* of politics and *technai*. Like him, several figures of *protoi heuretai*, inventors or discoverers of an object, phenomenon, or institution are mythological characters,[28] as demonstrated by Palamedes, the *inventor* of the letters of the alphabet, the calculation of the months, the game of draughts; Aristeus, the discoverer of hunting tools and bee farming; Aminocles of Corynthus, *protos heuretes* of the triremes; Erictonius, *protos heuretes* of the quadriga; Cecrops, who instituted monogamous marriage in Athens; Dracon and Lycurgus, who established laws, and so on. As each word has its *etymon*, each city its eponymous, each rite and custom their *aition*, so each invention has its *heuretes*.

The folk-tale – not only in Greece – is in itself strongly aetiological, because among its different functions it also aims at clarifying the motive for many facts, binding the present to the past, the origin to the actual state. A tendency towards aetiology exists from the very beginning also in the literature of Greece. Overall it was meant to offer, in works of diverse nature, an explanation for myths, cults,

27 See *Orphicorum Fragmenta*, ed. by Otto Kern (Berlin: Weidmann, 1922), F 140, 220, 224.

28 Adolf Kleingünter, *Protos Heuretes: Untersuchungen zur Geschichte einer Fragestellung* (Leipzig: Akademie Verlag, 1933).

natural phenomena, and specific customs, which were often clarified by resorting to lesser known mythical stories. [29]

Several aetiological myths are, for example, present in Pyndarus: they explain the origins of games or cities, thus giving them a sort of official religious consecration and tracing human deeds back to a divine origin. In the fifth-century tragedy, mainly in Euripides, the *aition* reaches a rationalistic and cultural connotation, foreshadowing in part the scholarly value and cultural refinement that it would acquire in the Alexandrine age. In comedy, instead, it is bent to the ends of parody and becomes a real comical theme: poets invent, for example, aetiological legends to justify comical situations, or they create paraetymologies to link a proper noun to a pejorative one.

Historians and philosophers also make use of similar themes: nevertheless, the *aition* becomes a widespread theme and an autonomous literary genre in its own right only during the Hellenistic age, when an interest in ancient times, combined with a taste for scholarship, leads to the idea of telling mythical stories not only for themselves, but also as explanations for human deeds.

The *Aitia* by Callimachus are without doubt the most representative. The main topic of his collection of elegies in four books is the origin, mostly mythical, of names, customs, traditions or cults. In Book I, for example, the origin of the Lindus cult of 'Heracles who feasted on an ox' is shown, during which the hero is celebrated with insults and with the offer of a non-dismembered ox. At Lindus, on the island of Rhodes, Heracles devoured an ox belonging to an old peasant who had insulted him, thus giving birth to the ritual. The tale of Linus then explains the Argive ritualistic killing of dogs found loose on the streets during the month of Arneios. The daughter of the Argos king, being afraid of Apollo, the father of her child, exposed him, but he was devoured by dogs: his death led to the origin of the myth. In Book II Callimachus deals with the origins and founders of Sicilian cities, whilst Books III and IV contain the *aitia* of rites, feasts and customs. Among them the famous *Lock of Berenice*, in which we trace the origin of a constellation from a curl offered to the gods by Queen Berenice to propitiate the return from the war of her husband Tholomeos III, who then disappears and appears in the sky.

29 Giancarla Codrignani, 'L'*aition* nella poesia greca prima di Callimaco', *Convivium*, n.s., 26 (1958), 227–45.

Alessandro Zironi

Searching for Origins: Indo-European Words and Nordic Representations

I. To the Origin of Meaning

What do we intend by saying 'origin'? How many words are connected with the meaning of 'origin'? Since words are deeply rooted in the culture of the speakers, it is impossible to suggest considerations embracing all languages. The choice to focus on the Indo-European languages is due to the extension of the territories in which Indo-European populations settled (from Iceland to south-western China), and to the antiquity of their civilization, which goes back at least to the fifth millennium BCE; possibly, following some recent theories, to 100,000 years BCE. [1] Different words used for 'origin' will thus be examined in relation to some representative Indo-European linguistic families, namely Indo-Iranian, Greek, Latin, and Germanic, and particular attention will be devoted to the semantic fields in which the words are collocated. The linguistic results will then be used in order to investigate a mediaeval Icelandic representation of the origin of the earth, which is connected with the reception of North. Both Old Norse poetry and contemporary artistic productions prove that Iceland has constantly been conceived of as a place which evokes origins.

II. How Indo-Europeans Say Origin

II.I Sanskrit

In Sanskrit there are several words connected with the idea of origin:
– *prabhavá*, 'the cause of existence', 'source', 'origin'. The verbal form is

1 Franco Cavazza, *Lezioni di indoeuropeistica con particolare riguardo alle lingue classiche (sanscrito, greco, latino, gotico)* (Pisa: Ets, 2001), pp. 99, 195–96.

prabhavati [2] 'to spring up, to appear, to become visible', and also 'to rule', 'to have the power over someone or something'. [3] The word is composed by the prefix *pra-* 'before' (see IE **pro-, prō*) [4] and the IE root **bheu-, bheuə, bhōu, bhū*, which originally should have meant 'to grow, to swell' (P, pp. 146–47) (see Greek *fuomai* 'to become, to grow') [5], but it became one of the roots of the athematic verb to be (as, for instance, in the infinitive in English). *Prabhavá* means 'origin as something that is grown before existing as a being, as a creature'. Therefore, the act of origin connected with this word is the moment or the will preceding the growth, the springing up of a creature.

- *utpatti*, 'arising, birth, origin'. It shares the same composition of the verb *utpatati* 'to rise, to originate' (MW, p. 180). Both are composed structures, constituted by the prefix *ut-* (the same as the English preposition *out*) (P, pp. 1103–04) and the root *pat* which refers to the IE root **pet- petə- ptē-, ptō* 'to fly, to fall', like the Homeric Greek *pétomai* 'to fly, to flee' (P, p. 825). Origin, in this sense, is the moment in which something or someone is going out of or arising from a body, from the earth, or from something which can contain. Also the expressions *udgā*, *udgata* have a similar sense: 'to rise, to begin' and 'gone up, risen, ascended' (MW, pp. 186–87), composed with the root that generates the English verb 'to go'; 'to go out' thus means that something or someone has its origin in the coming out.
- *mūla*, 'root of a plant or tree', but also 'origin, commencement, beginning' (MW, p. 826). The sense is clear. The word is composed by the IE root **mū* which means 'mouth' (P, p. 751). Consequently, *mūla* has the sense of 'mouth of the tree' and, in a metaphoric way, 'origin' as something that comes out from a mouth.
- *agra*, 'foremost, first, prominent, the beginning', also with a temporal sense (WM, p. 6). It comes from IE root **agro-*, 'top, peak, beginning' (P, p. 8). It is interesting to compare these meanings with the Latin personal name *Agrippa*, that could derive from a form **agri-p(e)d*, 'footling breech'(P, p. 9).
- *yóni*, 'womb, uterus, origin, spring, fountain' (WM, p. 858). It must be noticed that the IE root **jeu-* means 'to begin a movement' (P, p. 507). Consequently, the Sanskrit form has to do with a body that bears fruit, with a container from which something is to come out.

2 All Sanskrit verbal forms are indicated with the third person singular of the present tense. As the same person would sound unusual in English, I prefer to use the infinitive form.

3 Monier Monier-Williams, *A Sanskrit – English Dictionary Etymologically and Philologically arranged with special reference to cognate Indo-European Languages* (Dehli: Motilal Banarsidass Publishers, 1995), p. 684 (hereafter: MW).

4 Julius Pokorny, *Indogermanisches Etymologisches Wörterbuch*, 4th edn, 2 vols (Tübingen and Basel: A. Francke Verlag, 2002), I, p. 813 (hereafter: P).

5 The infinitive form is used also for the English translation of Greek verbs.

II.II. Greek

In Greek there are two principal forms conveying the sense of origin: *arhè* and *génos.*

- *arhè* is a substantive which means 'beginning, origin', but also 'government, dominion' and is a deverbal derivation of the verb *árhō* 'to begin, to dominate'. [6] Its etymological origin is unknown and disputed, as no proposed theory is completely convincing. Nevertheless, it is surely interesting to notice that the verb *árhō* has also the sense of 'to rise over someone, to dominate' and can be put in relation with *órhos*, which means 'branched', typically used for a vineyard. [7] The two forms could be apophonically related, as *órhos* should be the strong apophonic degree of *árhō*. [8] Even if the etymology of *arhè* is controversial, some light on this word is shed in a *Commentary on St. John's Gospel* by Origen of Alexandria, written in the third century. Analysing the very beginning of the Gospel, *en arhè 'o lógos* (John, I, 1), Origen declares that the word *arhè* has several meanings both in the Greek language and in the Bible. [9] The most meaningful ones can be summarized as follows:
  1) *arhè* is the passage from one place to another, the beginning of a process leading to the point of apocatastasis, that is the return to a primeval, pristine condition. [10] In other words: even if the process seems to lead forward, it actually goes back to a pristine moment or situation.
  2) the beginning is the origin (*génesis*), as it comes out in Old Testament (Job, 40, 19: 'this is the beginning of the material creation').
- What is essential in Origen's reflection on the word *arhè* is the semantic collocation of the term in a kinetic dimension, even if the word also continues to refer to the starting point. In other words, the idea of *arhè* is not static, is not locked in the tight mesh of a single moment. On the contrary, it interferes and goes together with the continuation of that temporal beginning. Finally, *arhè* shows possible phonological implications with the form *órhō*. There is a sufficient number of cases in Greek and Byzantine literature, both pagan and Christian, in which the birth of a great person (for instance a future king or a

6 Hjalmar Frisk, *Griechisches Etymologisches Wörterbuch*, 2nd edn, 3 vols (Heidelberg: Carl Winter Universitätsverlag, 1973), I: A – Ko, pp. 158–59 (hereafter: F).

7 Emile Boisacq, *Dictionnaire étymologique de la langue grecque* (Heidelberg, Carl Winter, 1923), p. 85.

8 Eduard Schwyzer, *Griechische Grammatik im Anschluss an Karl Brugmanns griechische Grammatik. Allgemeiner Teil und Lautlehre* (München: C. H. Beck'sche Verlagsbuchhandlung, 1934), p. 685.

9 Origene [Origen], *Commento al Vangelo di Giovanni*, ed. by Eugenio Orsini (Torino: UTET, 1968), I, XVI, 90, p. 143.

10 Domenico Pazzini, *In principio era il Logos. Origene e il prologo del Vangelo di Giovanni* (Brescia: Paideia, 1983), p. 39.

saint) is foretold by the vision of a tree which grows out of a female body. In many of these cases the tree is a vine tree. [11] Therefore, *arhè* and *órhō* are perhaps more interconnected than it could seem at first glance. Consequently, words such as tree, dominion, beginning and origin seem to be interrelated.

- *génos*. It derives from the verb *gígnomai* 'to bear, to become' (F, pp. 297, 307 – 08). It is a very old Greek verb that can be compared with to IE root **gen* (P, pp. 373 – 74). The word is near to the sense of *arhè* where the latter means 'beginning', but it differs from *arhè*, as it testifies more precisely to the sense of birth, of beginning of a life, also from a typological point of view. The use of *génos* allows for identifying race, genre, family as a uniform group of living beings or inanimate things, without implying a kinetic denotation: *génos* is static, rigid and in this sense it is a sort of counterpart of *arhè*.

II.III. Latin

Obviously, the Latin word is *orīgo*, a deverbal noun coming from *orior*, which means 'to raise, to come up, to bear'. [12] It goes back to an IE root **er*-, **or*- **r*- meaning 'to set out, to excite, to provoke' (P, pp. 326 – 30), and it can also be found in the Greek *érnos* 'bud, sprout, shoot'. [13] Correspondences are also possible with the Germanic languages, where the word is connected with trees and their growth: OIcel. *rinna*, 'to grow, to come up', [14] OHG *rono* 'base of a tree'. [15] Otherwise, in the Germanic languages, the IE root **er*-, **or*-, **r*-, in this case from the form **re-nu-ō*, implies the idea of running, above all with reference to liquid substances: OIcel. *rinna* 'to flow, to run', OHG, OS, OE *rinnan* 'to flow, to run'. Therefore, Latin and the Germanic languages have preserved the primeval meaning of the IE root, expressing the growth of trees, flowing waters, etc., and from there derives the meaning 'origin', as something or someone that sprang up or flew. The Romance languages have carried on the Latin word, while the Germanic languages (with the exception of English which received the word from Old French) have discarded it. There is only the noteworthy exception of

11 A list of passages can be recollected in Alessandro Zironi, 'Il sogno premonitore dell'albero nella tradizione germanica medievale', in *Filologia romanza, filologia germanica: intersezioni e diffrazioni*, ed. by Anna Maria Babbi and Adele Cipolla (Verona: Fiorini, 1997), pp. 343 – 65 (pp. 349 – 51).

12 *Thesaurus Linguae Latinae*, IX, 2 fasc. VII (Leipzig: Teubner, 1990), coll. 981 – 1007.

13 Pierre Chantraine, *Dictionnaire étymologique de la langue grecque. Histoire des mots* (Paris: Klincksieck, 1980), p. 374.

14 Richard Cleasby, Gudbrand Vigfusson, and William A. Craigie, *An Icelandic – English Dictionary*, 2nd edn (Oxford: The Clarendon Press, 1962), p. 493 (hereafter: CVC).

15 Rudolf Schützeichel, *Althochdeutsches Wörterbuch*, 4th edn (Tübingen: Max Niemeyer, 1989), p. 214.

Icelandic, which uses the composed word *uppruni* to mean 'origin' (CVC, p. 656): *uppruni* is something that runs up, that grows in a vertical way, as trees effectively do.

II.IV. Germanic

The most ancient Germanic word for 'origin' is the Visigothic **frum*, transmitted through the fourth-century translation of the Bible. [16] We encounter twice – in John 15,27 and John 16,24 – the expression *fram fruma*, translating the Gospel's text *ap' (ex) arhēs*. Greek *arhè* is therefore Gothic **frum*. This fact demonstrates two issues:

1) the Germanic languages and Latin do not have a word meaning 'origin, beginning' which can be etymologically related to Greek *arhè*. Actually, *arhè* could be a non-Indo-European word.
2) Gothic **frum* and Greek *arhè* have substantially the same meaning. We cannot doubt it, as the Gothic translator was very acute in every part of his work and, moreover, he was surely a perfectly fluent bilingual speaker of Greek and Gothic by birth.

Gothic **frum* comes from the IE root **prmo-* (from which the Latin *primus* derives) (P, 813 – 14). The same word appears in Old English as *fruma* and *frymð*, also in the Gospel's passages in which the term is *arhè* in Greek and *orīgo* or *principium* in Latin (from which the Anglo-Saxons translated the Gospels in the tenth and eleventh centuries). For Germanic people, therefore, the origin is the beginning, it marks the very first moment or accident from which the other things (or living beings) descend. It defines the starting point: there is no interest in growing, flowing and, consequently, in the kinetic idea that emerges from *orīgo* or *arhè*.

The idea of growing or movement is given in Germanic by the IE root **sprengh-* (P, p. 998), the same of the English verb 'to spring'. It must be observed that the German word for 'origin' is now *Ursprung*, which refers to the point of beginning or the moment from which something or someone has begun to rise, to grow.

It is possible to argue that the Germanic languages need two forms in order to express the concept of 'origin':

1) a word which focuses on the starting point;
2) a word which highlights the kinetic movement from a starting point. Origin, in this way, is perceived in its continuous springing and growth: it is not a

16 Winfred P. Lehmann, *A Gothic Etymological Dictionary* (Leiden: E. J. Brill, 1986), p. 129.

moment of the past but something present, because it has been springing or growing.

II.V. The Indo-European Lexicon for Origin

The study of Indo-European linguistic families shows that the idea of origin is explained in two different ways:

1) Origin is described as 'growth, springing'. It is not important to identify or focus on the exact moment in which something began; on the contrary, the Indo-Europeans look at the development produced by that beginning, as they are more interested in the kinesis that characterizes the thing or the living being. No difference emerges between stones and trees, animals and human beings. To this conception words as Sanskrit *prabhavá*, *utpatti*, Greek *arhè*, Latin *orīgo*, Icelandic *uppruni*, German *Ursprung* can be referred. The key-word explaining this idea of origin is 'to spring': in a fountain we do not pay attention to the exact point from which water springs out, but to the water's jet and its route in the air.
2) The other area covered by a lexicon for origin is more interconnected with the place or the moment in which something or someone has its beginning: Sanskrit *mūla*, *apra*, *yóni*, Greek *génos*, Gothic **frum*, Latin *primus* have to do with the place where someone is born, with the coming out of something or the idea of being the point of departure. Calling again to our mind the image of a fountain, we are now looking at the mouth from which water goes out, as we are not interested in water itself. To this idea English words such as *pristine* and *primeval* refer, being very different in meaning from *origin*. Thus, primeval or pristine are not synonyms of origin: origin shows the movement, pristine the stasis.

III. *Ginnungagap*

The contrast between stasis and kinesis connected with origins can be gathered from *Völuspá*, the cosmogonic poem which opens *Edda*, a collection of Nordic mythological and heroic poems. A prophetess born before the gods narrates to Wodan, the Head of the gods, the origin of existence. The god is trying to obtain from the seer not only prophetic knowledge about the earliest past but also the capability of foretelling the future. Wodan's request is only a literary expedient in order to explain the origin of the earth and cosmos and their final destiny. *Völuspá* was conceived at the very end of the tenth century, or at the beginning of the eleventh century. Nobody knows if the poem had an original oral composition, but surely it was offered for oral circulation. Owing to the different directions the text took, it has been transmitted in six manuscripts copied in different centuries. Some Christian references in the poem have fuelled the hypothesis that it was composed after the christening of Iceland, which happened in the year 1000. Nevertheless, it has been pointed out that the millenarianism which characterizes some strophes can be connected with a religious initiative carried on in Nordic countries by the Norwegian king Ólafr Tryggvason during the latest part of the tenth century. [17] The most ancient copy of *Völuspá* was presumably transcribed about the year 1270 in the so-called *Codex Regius* 2365 4to. In that manuscript the poem is composed by 66 strophes organized in a sort of tripartite structure. The first section focuses on the creation of earth and firmament; the second deals with the narration of the world, the gods and man; the third and most famous section is about the *ragnarrök*, 'the destinies of the gods', the pagan end of the world. The poem is concluded by strophes related to a new birth, a new creation. The third strophe is particularly interesting:

Ár var alda. Þat er Ymir bygði,
vara sandr né sær né svalar unnir;
iorð fannz æva né upphiminn,
gap var ginnunga, enn gras hvergi [18]

In earliest times did Ymir live:
Was not sea nor land nor salty waves,
Neither earth was there nor upper heaven,
But a gaping nothing, and green things nowhere. [19]

The expression *Ár var alda* can be also translated 'At the beginning of times', the moment of the origin of creation. In the following lines a creation *ex nihilo* is actually presented, as was the case in Christian tradition: a series of negations (*né*) excluded the existence of sea, waves, land, heaven, grass and trees. [20] Only a place existed, *ginnungagap*. The *gap [...] ginnunga* that appears in the seventh line is unanimously considered coincident with the form *ginnungagap* transmitted in the *Gylfaginning* by Snorri Sturluson, a sort of poetic manual written during the twelfth century. In *Gylfaginning*, *ginnungagap* is described as follows:

17 *Völuspá. Un'apocalisse norrena*, ed. by Marcello Meli (Roma: Carocci, 2008), p. 17.
18 *Edda. Die Lieder des Codex Regius nebst verwandten Denkmälern*, ed. by Gustav Neckel and Hans Kuhn, 5th edn (Heidelberg: Carl Winter Universitätsverlag, 1983), *Völuspá*, III, 1–8, p. 1.
19 *The Poetic Edda*, trans. by Lee M. Hollander (Austin: University of Texas Press, 1962), p. 2.
20 *Völuspá*, pp. 60–61.

Ginnungagap, the part that faces in a northerly direction, was filled with the weight and heaviness of ice and rime and there was vapour and a blowing inwards from it. But the southerly part of Ginnungagap cleared up in the face of the sparks and molten particles that came flying out of the world of Muspell. [21]

Snorri Sturluson is explaining that there were two material entities. The first one, *Niflheimr*, was collocated in the North. It is a place where there is a source, *hvergelmir*, from which hot waters (*hverr*, 'termal water' in Icelandic) come out, forming eleven rivers. These rivers take with them material that, flooding away from the hot source, becomes solid and cold, covered by ice and frost. At the other side, in the South, Snorri Sturluson collocates *Muspellheimr*, a place from which sparks come out. Iced material and sparks arrive at an intermediate area, between North and South, which is called *Ginnungagap*. *Ginnungagap* is the melting area of two opposite materials and two opposite temperatures, but what it is exactly remains vague. There are two main theories about the interpretation of this composed noun. The first one sees *Ginnungagap* as a metaphoric expression, a *kenning*, for 'the hawk's cavity': Old Norse *ginnungr* also means 'hawk', and *gap* is cavity. The cavity where the hawk can fly is obviously the sky, contained by the heavenly vault. [22] In this case, the poem *Völuspá* would lead us to believe that only the air existed between *Niflheimr* in the North and *Muspellheimr* in the South. The second interpretation of *Ginnungagap* compares Old Norse *ginnunga*, a plural genitive form, to the Old High German *ginunga*, a word that glosses the Latin *hiatus*, *rictus*, which is a carsic cavity in the ground. [23] Old High German and Old English also have the verbs *ginen* and *ginian* respectively, meaning 'to open wide'. These linguistic and semantic correspondences between Old High German and Old English cannot be easily discarded. Moreover, in the eleventh-century work *Gesta Hammarburgensis Ecclesiae Pontificum* (*Deeds of Bishops of the Hamburg Church*) by Adam of Bremen, a *scholium* to the text alludes to the northern part of the Atlantic ocean beyond Iceland, the *Ultima Thule*, where there is a *ghimmendegop*, an abyss in which waters fall:

Qui latitudinem septentrionalis oceani perscrutatus navibus tandem caligantibus ante ora deficiens mundi finibus inmane baratrum abyssi retroactis vestigis pene vix salvus evasit. [Istud eciam dixit quidam notabilis Carthusiensis presentium scriptori, et est verum. Dicitur istelocus in eorum ydyomate Ghimmendegop.

21 Snorri Sturluson, *Edda*, trans. and ed. by Anthony Faulkes (London: Everyman, 1997), p. 10.

22 *Völuspá*, pp. 61–62.

23 *The Poetic Edda*, ed. by Ursula Dronke, 2 vols (Oxford: The Clarendon Press, 1997), II. Mythological Poems, pp. 112–13.

Miles vero capitaneus regis dicebatur Olyden Helghesson, nauta vero Gunnar Raswen.] (Scholium d66, mss. A3a, A3a') [24]

The Indo-European languages imagine origin as the moment in which something comes out, or like a womb, a cavity where life begins. *Ginnungagap* is thus imagined as a place filled by all elements necessary to life, hot and cold, moist and dry. Origin is consequently the abyss, the cleft, the gap, where the void is finally substituted by the creation of the earth. Ice and hot waters which enable creation remind of the natural Icelandic environment, where the primeval condition of the world is still visible.

Ursula Dronke stressed that *ginunga* should be the heathen Germanic word for chaos. [25] Icelanders reinforced this term by adding the tautological expression *gap*, a sort of gloss to *ginnunga*: [26] chaos and yawning emptiness. This narration has parallels in Indo-European culture: it could be sufficient to compare it to Ovid's *Metamorphoses* [I, 430 – 37] and to the cosmogonic Zoroastrian text *Bundahisn*. [27] This last text, written in Persian in the eighth or ninth century, has been compared to the embryologic theories, in particular the ones in which the union of female and male gametes creates a zygote. [28]

As a sort of provocation, it is interesting here to remark that there are two kinds of gametes: the first one, the male gamete, is more movable, kinetic; the female gamete is more static. Perhaps it is not by chance that the words denoting 'origin' in a static way are connected to the female reproductive apparatus, while the kinetic words for 'origin' also enclose the idea of a vertical growth, which can also be interpreted in a symbolic or totemic way.

In the abyss of *Ginnungagap*, the meeting of the two parts, one hot and dry, the other cold and moist, produces the filling of the gaping maw and the creation of the first being, Ymir. The primeval creature, the giant Ymir, will be thrown into *Ginnungagap* after having been killed by three gods, Oðinn, Vili and Vé. The corpse of Ymir is then dismembered by them. The correspondence with the Indo-European tradition is evident: the divine triad substitutes a previous belief, the dismembering of a corpse is the founding moment of creation. From the corpse of Ymir, we learn from various Old Norse texts, the world draws all its elements: blood → sea; flesh → land; bones → mountains; teeth → stones; hair

24 Adamo di Brema [Adam of Bremen], *Storia degli arcivescovi della Chiesa di Amburgo*, ed. by Ileana Pagani (Torino: UTET, 1996), p. 498.
25 *The Poetic Edda*, ed. by Ursula Dronke, p. 113.
26 Ibid., p. 112.
27 Bruce Lincoln, 'The Center of the World and the Origins of Life', *History of Religions*, 40 (2001), pp. 311 – 26 (p. 312).
28 Clarisse Herrenschmidt, 'Entre Perses et Grecs. I. Démocrite et le mazdéisme: Religion, philosophie, science', *Transeuphratène*, 11 (1996), 115 – 43.

→ trees; brain → clouds; skull → vault of heaven. The earth is actually created in *Ginnungagap*, the place where the oppositions are solved. Cosmos and earth have consequently a double syncretistic origin: an empty place which becomes the meeting point of opposite substances and conditions. After blending these materials, the earth and heavenly vault are created from the corpse of Ymir by the divine triad.

IV. Iceland, the Craters, the Abyss

Iceland has always been considered primordial and primitive in its nature. One of the most famous writers of the Nordic Middle Ages, Saxo Grammaticus, who lived at the end of the eleventh century, wrote a historical work, *Gesta Danorum*, about the kings of Denmark. Like the majority of historical works of late Antiquity and Middle Ages, also the *Gesta Danorum* begin with a geographical preface in which the places where the subsequent events will happen are described. The description of Iceland is particularly interesting:

> On the west of Norway comes the island called Iceland, with the mighty ocean washing round it: a land very squalid to dwell in, but noteworthy for marvels, both strange occurrences and objects that pass belief. A spring is there which, by the malignant reek of its water, destroys the original nature of anything whatsoever. Indeed, all that is sprinkled with the breath of its vapour is changed into the hardness of stone. It remains a doubt whether it be more marvellous or more perilous, that soft and flowing water should be invested with such a stiffness, as by a sudden change to transmute into the nature of stone whatsoever is put to it and drenched with its reeking fume, nought but the shape surviving. [...] In this island there is likewise a mountain, whose floods of incessant fire make it look like a glowing rock, and which, by belching out flames, keeps its crests in an everlasting blaze. [...] a land lying close to the extreme of cold can have such abundance of matter to keep up the heat, as to furnish eternal fires with unseen fuel, and supply an endless provocative to feed the burning. [29]

The volcanic lapilli and the siliceous hot waters which cause the formations of rocks are compared to the two elements, the first coming from *Muspellheimr*, the second from *Niflheimr*, that meet in *Ginnungagap*. Iceland seems to be the place of *Ginnungagap*, the actual place of the origins.

Saxo's image of the island does not derive from insular writers, who never described Iceland as an impervious or infernal place, but it reflects the common

29 Saxo Grammaticus, *The First Nine Books of the Danish History*, trans. by Oliver Elton (London: David Nutt, 1894), Preface, p. 10. See also http://omacl.org/DanishHistory/preface.html) [accessed 3 September 2009].

idea of Iceland as a primeval country. This statement, which was born outside the island, became more and more vigorous through the centuries.

V. Jules Verne: Origin from the Abyss

Although some interesting publications appeared during the modern age in Scandinavia and France, the interest in Iceland increased towards the end of the eighteenth century and in the nineteenth century, when travelogues about Iceland became copious and constituted a literary sub-genre.

The obsession with the knowledge of Iceland was particularly strong in Victorian Britain, but it also spread to the other side of the Channel. [30] In 1861, Jules Verne voyaged through Scandinavia, visiting only Denmark and Norway. Three years later his second novel, *Voyage au Centre de la Terre*, appeared in France. The protagonist, Otto Lidenbrock, a mineralogist and professor in Hamburg, casually discovered in a manuscript of Snorri Sturluson's *Heimskringla* a leaflet written by the seventeenth-century Icelandic alchemist Arne Saknussemm. It is a cryptogram in runic characters revealing the point from which the centre of the earth can be reached. Professor Lidenbrock, his nephew Axel and the Icelandic servant Hans will reach the centre of the earth beginning their voyage from a crater situated in the Icelandic glacier Snæfellsjökull. They will emerge two months later from the Mediterranean volcano Stromboli. The impressive massif of the Icelandic volcano surmounted by the glacier is the place where the abyss is open. In the novel, the young Axel Lidenbrock declares his deep attraction towards the abyss:

> Now began our real journey. [...] I had not yet ventured to look down the bottomless pit which I was about to take a plunge. [...] I approached the central chimney. I have already mentioned that it was a hundred feet in diameter, and three hundred feet round I bent over a projecting rock and gazed down. My hair stood on end with terror. The bewildering feeling of vacuity laid upon me. I felt my centre of gravity shifting its place, and giddiness mounting into my brain like drunkenness. There is nothing more treacherous than this attraction down deep abysses. [31]

Why must the centre of the earth be searched for starting from Iceland? The abyss over which the Snæfellsjökull crater leans out, the way that leads to the

30 Andrew Wawn, *The Vikings and the Victorians. Inventing the Old North in Nineteenth-century* Britain (Cambridge: Brewer, 2000), pp. 283 – 311.

31 Jules Verne, *Journey into the Interior of the Earth*, trans. by Frederick Amadeus Malleson (London: Ward Lock & Co., 1877), chapter 17, <http://jv.gilead.org.il/wolcott/CE-allc/17.html> [accessed 3 September 2009].

heart of the globe could not be placed in another country. It is not only a question of reminiscences caused by the recent travel of Jules Verne to Scandinavia. At the end of the novel, in the second-last chapter, Axel sets Stromboli against Iceland:

> What a journey we had accomplished! How marvellous! Having entered by one volcano, we had issued out of another more than two thousand miles from Snæfell and from that barren, far-away Iceland! The strange chances of our expedition had carried us into the heart of the fairest region in the world. We had exchanged the bleak regions of perpetual snow and of impenetrable barriers of ice for those of brightness and 'the rich hues of all glorious things'. We had left over our heads the murky sky and cold fogs of the frigid zone to revel under the azure sky of Italy! [32]

The loneliness of Iceland and its grey landscape are here opposed to the luxuriant Mediterranean island. Origins are in a certain sense connected with absence of life. Where flora can be contemplated, it is not possible to imagine origins, because they are to be found where nothing exists before the beginning. Origins are at the same time the vacuum, the abyss. Iceland in European culture answers the need to identify a primeval place where the origins of the world can be imagined.

VI. Artistic Representations of Primeval Iceland

The penetration into the earth from the place where the landscape reveals its pristine condition and the way into the volcano symbolically intended as a sort of back-journey into the uterus where the earth was generated are in my opinion also expressed in *Cave* (2003), a short film by the Finnish artist Salla Trykkä. Trykkä puts at the centre of her narration a woman who leaves a house and enters a cave. After having plunged her fingers into a little puddle of water blended with grey dust, she finally runs out from the cave through another mouth, different from the one she penetrated. The landscape in which she finds herself is rocky, probably lavic: maybe Iceland? [33] We can compare this short film with the photograph taken the same year by the Italian artist Walter Niedermayr, *Leihrnjukur I*. Leihrnjukur, a lava field in the Krafla volcanic area located in the north-eastern part of Iceland, was generated in 1984. [34] Consequently, the surface is still thin, the air is sulphurous; in few words primordial, primeval. The

32 Ibid., chapter 45, http://jv.gilead.org.il/wolcott/CE_allc/45.htm [accessed 3 September 2009].

33 *Northern Lights. Reflecting with Images*, ed. by Filippo Maggia and Silvia Ferrari (Cinisello Balsamo: Silvana Editoriale, 2007), pp. 27 – 33.

34 Ibid., pp. 70 – 71.

same idea of Iceland as a primeval land which overcomes the human being also emerges from the recent film by the German artist John Bock, *Skipholt* (2005).[35] For Trykkä, Niedermayr and Bock the rocky and volcanic landscape is the representation of a world in which man is marginal and insignificant, a little figure contemplating the vastness that has been filled by new rocks created by the meeting of incandescent lava and cold air.

VII. Apollo 11 and Pierre Huyghe's *One Million Kingdoms*

The representations of origins from *Völuspá* to current artworks seems to converge in the Icelandic landscape. Interestingly enough, scientists have considered Iceland as a privileged place for investigating the origins. It is here necessary to recall the first travellers who went to Iceland at the beginning of the nineteenth century, driven by the desire to find uncorrupted and, consequently, original material for the development of scientific research.[36] It would be superficial to consider such an attitude as a perception of the world resulting from a romantic sensibility. In the twentieth century, at the end of the sixties, the NASA Apollo program contemplated the landing of the man on the lunar ground. A series of tests for equipments, LEM and astronauts were planned. The place that was considered congenial for those scientific simulations was the Eldhraun lava field in South Iceland. Its name means 'fire-lava':[37] it is surely a matter of chance, but nevertheless it is surprising that fire and solidified rocks are again melted in a place-name, as they were in the original fusion of elements in *Ginnungagap*. Eldhraun, a meaningful name for a cultural meditation, was chosen by scientists as the place for a new beginning, for the adventure of man on the lunar surface.

The implications between Iceland and the Apollo mission were explored by the French artist Pierre Huyghe in his short animated film *One Million Kingdoms* (2001). The core element is a manga character named Annlee. The film begins with an image which evokes the 1865 descent from Snæfellsjökull to the centre of the earth described by Jules Verne, then the moon test area collocated at the left of Snæfellsjökull. A vertical line joins Iceland with a dark moon, next to which 1969, the year of the landing of Apollo 11, is indicated. Finally, at the level of the equator, the words 'expedition departure 1998' are reported and a line connects this point with Iceland on the top. Then the animation begins. Annlee, the

35 John Bock, *Skipholt* (Köln: Verlag der Buchhandlung Walther König, 2008); Marco De Michelis, *Il sublime è ora / The Sublime is Now* (Milano: Skira, 2008), p. 37.

36 Wawn, pp. 283–311.

37 Ennio Cavalli, *Il divano del Nord. Viaggio in Scandinavia* (Milano: Feltrinelli, 2005), pp. 81–82.

manga-character, walks in a lunar landscape while an off screen voice is speaking. Only afterwards does one perceive that it is the same Annlee who is speaking. The voice is a rearrangement of Neil Armstrong's recording during the Apollo 11 expedition. The voice of Armstrong is the meta-structural component which connects Jules Verne, Icelandic tests and lunar expedition:

The *Journey to the Centre of the Earth* by Jules Verne begins in Iceland in the Snoefells Jokull crater at the north of the island. The conquest of space starts there on that desert of lava. The first images of Neil Armstrong hopping in his space suit in the middle of a desolated landscape were shot there. This is an expedition through territories that are topologically similar. [38]

The description of a desolate landscape is reinforced by Armstrong's voice remarking that the tests in Iceland prepared mankind for the 'spectacle of desolation' the moon would offer:

The conquest of space, which was a dream until now [...] had become an illusion. We want to enter the unknown, when the greatest mysteries are right here [...] Here under our steps [...]. [39]

The disillusion towards lunar expeditions and the dismissing of lunar exploration projects with the Apollo 17 in 1972 are evoked in the words chosen by Pierre Huyghe. While Annlee walks, the mountains around her peak and collapse following Armstrong's voice. We immediately understand that the landscape is the graphic realization of the sound-waves of the voice. The narration is intermingled with words from the real record of the lunar landing, but the final part is taken from the seventeenth chapter of *Voyage au Centre de la Terre* by Verne:

The real journey was beginning.
Fearful dangers lurked. I had not yet looked down into the bottomless pit which I was about to plunge into.
I was ashamed to step away. Now the time had come.
I leaned over a projecting rock and looked down [...] My hair stood on end [...]. The fascination of the void took hold of me. I felt my centre of gravity moving, and vertigo rising to my head like intoxication.
There is nothing more powerful than this attraction of the abyss. [40]

38 Pierre Huyghe, 'One Million Kingdoms (2001)', in Carolyn Christov-Bakargiev, *Pierre Huyghe* (Skira: Genève-Milano, 2004), p. 359.
39 Ibid., ivi.
40 Ibid., ivi.

VII. Searching for Origins

The conclusion of the animated film reports the desolated reflection of Pierre Huyghe about human solitude faced with the uncertainty of what surrounds the earth. Lunar expeditions gave no answer to the loneliness of man in creation and were also unable to explain the origin of existence. Only the human voice pronouncing the name of the narrator is audible in the void and unknown space:

> Nothing [...] with the possible exception of [...] the state of weightlessness.
> For a moment I was afraid that their words might be my own, brought back to me by an echo.
> I listened once more, and this time I clearly heard my name thrown through the space.[41]

Like in the *Völuspá*, nothing exists except the void that preceded creation. The search for origins is a way to make up for that void, those abysses which from the craters of Iceland have been transferred to cosmic dimensions and proportions: thanks to scientific progress, men have enlarged the boundaries of their celestial vault, but not solved, as yet, the need to find out the starting point, the origin of all.

41 Ibid., ivi.

Marusca Francini

Origo Gentis and Cultural Memory. The Origins of England as an Emblematic Case Study

I. Origins and Cultural Identity

The quest for origins is an anthropological standard feature in human cultures, since it is believed that origins can explain the present. Such a quest serves an ideological purpose: the sense of belonging to an ethnic group is rooted in the group's own history, and, above all, in the common shared origins, which strengthen its identity. [1] Collective memory defines the identity of a given group, since it creates a sense of common belonging; memory (the reference to the past), identity, and cultural perpetuation (the making of a tradition) are thus closely connected. [2] Memory not only builds the past, it also organizes the present and the future, since ethnic groups typically base their self-awareness on past events. [3]

By answering the question 'who are we?', narrations of origins aim at self-definition and consolidation of identity. They help stabilize social order, therefore they arise above all in periods of changes. The transition from Late Antiquity to the Middle Ages is particularly interesting: the Germanic peoples found their own kingdoms in the territories of the Roman Empire and were converted to Christianity. The Christian missionaries introduced the culture of writing; literacy brought the possibility of transmitting accurate written records, and the Germanic *gentes* wrote down their own history. In these historical works we can find a series of accounts, the so-called *origines gentium*, which relate the origins of a people: they describe the migration from the ancient homeland, the arrival and the vicissitudes of the ultimate settlement in a new territory. These accounts of the origins aim at defining the identity of the group and legitimizing the new social and political order which emerges from the Germanic migrations.

1 See Anthony D. Smith, *The Ethnic Revival* (Cambridge: Cambridge University Press, 1982).
2 See Jan Assmann, *Das kulturelle Gedächtnis. Schrift, Erinnerung und politische Identität in frühen Hochkulturen* (München: Beck, 1992).
3 See Maurice Halbwach, *La mémoire collective* (Paris: Presses universitaires de France, 1968).

In these accounts, after the migration the group clashes with other peoples, and these hostile encounters stress the unity and the qualities of the group. So, antagonism also contributes to the definition of collective identity.

There is a close link between historiography, which aims at a rational knowledge of history, and epos, which is the symbolic transfiguration of history. The most meaningful epic forms are handed down in historical literature; among them are the accounts of the origins and the deeds of the founding heroes of a people. The *origo gentis* is a kind of literary genre, embedded in many medieval historical works that relate the past of a *gens*; it represents the beginning of ethnogenesis, the formation of a people. According to the theory of ethnogenesis formulated by Reinhard Wenskus, [4] the *origo gentis* is exploited in order to give a sense of common identity to a number of originally separated groups at the moment they become a single people, characterized by common traditions. Ethnic affiliation is not simply biological: cultural and linguistic elements play a decisive role. Accounts of origins are part of these cultural elements; they were preserved by oral tradition well into the Christian period, when they were written down in historical works concerning the past of the Germanic peoples. [5] So, both poetic oral tradition and written historiography play an important role in the process of definition of the collective identity of a people.

A fundamental event in the cultural memory of the Anglo-Saxons is their invasion of Britain. The traditions pertaining to this event testify to the importance of the accounts of origins in the making of the Anglo-Saxon identity.

II. *Historia Ecclesiastica Gentis Anglorum* and the *Anglo-Saxon Chronicle*: the Origins of England

The Germanic colonization of Britain began in the fifth century, when the tribes of the Angles, Saxons and Jutes left the north-western coasts of the continent and settled on the island. The original inhabitants, the Celtic peoples of Britain, were forced to yield ground before the ever-increasing pressure of the invaders, and were driven westward into the regions of Cornwall and Wales, along the Atlantic coast. Two major historical works written in Anglo-Saxon England relate the beginnings of the Germanic settlement: *Historia Ecclesiastica Gentis Anglorum*

4 Reinhard Wenskus, *Stammesbildung und Verfassung. Das Werden der frühmittelalterlichen Gentes* (Köln – Wien: Bohlau, 1961).

5 In the Middle Ages a number of historical works relating the past of the Germanic *gentes* were written down: Jordanes wrote the history of the Ostrogoths; Gregory of Tours and Fredegar narrated the history of the Franks, Paul the Deacon wrote the history of the Lombards, and Widukind of Corvey that of the Saxons.

(*Ecclesiastical History of the English People*), composed by Bede in the first half of the eighth century, and the *Anglo-Saxon Chronicle*, a set of vernacular annals dating from the end of the ninth century. [6] The memory of the Anglo-Saxons' original migration is the foundation myth of Anglo-Saxon England; the *Historia Ecclesiastica Gentis Anglorum* and the *Anglo-Saxon Chronicle* present different pictures of this cultural myth and their different approaches will be here analysed and compared.

III. *Historia Ecclesiastica Gentis Anglorum*: the Anglo-Saxon Invaders as the Chosen People

Bede lived in the north of England, in the Anglo-Saxon kingdom of Northumbria, between the end of the seventh and the first half of the eighth century. He was a monk in one of the best known monasteries in the north, the twin foundation of Wearmouth and Jarrow. Bede composed homilies and school books, wrote on theology, hagiography, science, chronology, and history. [7] *Historia Ecclesiastica Gentis Anglorum* is his major historical work, completed in 731/732. In it Bede traces the gradual conversion of his people to Christianity. The great reputation and circulation of the book in the Middle Ages is demonstrated by the large number of medieval copies that survive: over one hundred and fifty. [8] The work, originally written in Latin, was translated into Old English during King Alfred's reign (ninth century); it was also borrowed and extracted from by chroniclers (even by the *Anglo-Saxon Chronicle*) and hagiographers, both in England and on the continent. [9]

The *Historia Ecclesiastica*, in five books, is arranged in a chronological order. This order is broken, to some extent, by the inclusion of saints' lives and accounts of miracles. [10] For the chronological arrangement of the work the author also exploited his knowledge of the computus; Bede had, in fact, written two treatises on the computation of time: *De Temporibus* (703), and *De Temporum Ratione* (725). The opening section of the first book outlines the background and

6 The Anglo-Saxon conquest is also related in *De Excidio Britanniae* (sixth century) by the Celtic writer Gildas, and in the *Historia Brittonum* (ninth century); on these works (which relate the events from the point of view of the Britons) and on Bede's, see Magali Coumert, *Origines des Peuples. Les Récits du Haut Moyen Âge Occidental (550–850)* (Paris: Institut d'Études Augustiniennes, 2007).

7 A list of his works was appended by Bede to the last chapter of *Historia Ecclesiastica*.

8 Antonia Gransden, *Historical Writing in England c. 550 to c. 1307* (London: Routledge & Kegan Paul, 1974), p. 16.

9 Ibid., p. 17.

10 Ibid., p. 16.

the context of the conversion of the Anglo-Saxon people to Christianity. This history proper begins in chapter 23 with the arrival in Kent of the Roman mission sent by Pope Gregory the Great and led by Augustine. In the introductory part we find an account of the arrival of the Anglo-Saxons in Britain and their occupation of the land, the so-called *adventus Saxonum*. The first book begins with a geographical description of Britain, drawing on authorities such as Pliny, Orosius, and Gildas. Britain is described as a good and fertile land (*opima*) – a kind of Promised Land. The mention of the vineyards of southern Britain is not found in the Latin sources used here by Bede; it springs from the author's own experience and it aims at emphasizing the depiction of Britain as the Promised Land. Bede goes on to discuss the earliest inhabitants of the island and gives a brief account of the Roman occupation; at this point, in chapters 15 and 16 he proceeds to narrate the story of the Anglo-Saxons' invasion.

Ch. 15: In the year of our Lord 499 Martian became Emperor with Valentinian [...]. In his time the Angles or Saxons came to Britain at the invitation of King Vortigern in three long-ships, and were granted land in the eastern part of the island on condition that they protected the country: nevertheless, their real intention was to subdue it. They engaged the enemy advancing from the north, and having defeated them, sent back news of their success to their homeland, adding that the country was fertile and the Britons cowardly. Whereupon a larger fleet quickly came over with a great body of warriors, which, then, when joined to the original forces, constituted an invincible army. These also received from the Britons grants of land where they could settle among them on condition that they maintained the peace and security of the island against all enemies in return for regular pay.
These new-comers were from the three most formidable races of Germany, the Saxons, Angles, and Jutes [...].
It was not long before such hordes of these alien people vied together to crowd into the island that the natives who had invited them began to live in terror. [...] the Angles [...] began to turn their arms against their allies. They began by demanding a greater supply of provisions; then, seeking to provoke a quarrel, threatened that unless larger supplies were forthcoming, they would terminate their treaty and ravage the whole island. [...] In short, the fires kindled by the pagans proved to be God's just punishment on the sins of the nation, just as the fires once kindled by the Chaldeans destroyed the walls and buildings of Jerusalem. For, as the just Judge ordained, these heathen conquerors devastated the surrounding cities and countryside, extended the conflagration from the eastern to the western shores without opposition and established a stranglehold over nearly all the doomed island. Public and private buildings were razed; priests were slain at the altar, bishops and people alike, regardless of rank, were destroyed with fire and sword [...].

Ch. 16: Their leader [of the Britons] at this time was Ambrosius Aurelianus, a man of good character and the sole survivor of Roman race from the catastrophe [...] Under his leadership the Britons took up arms, challenged their conquerors to battle, and with God's help inflicted a defeat on them. Thenceforward victory swung first to one side and then to another, until the battle of Badon Hill, when the Britons made a considerable slaughter of the invaders. [11]

The Northumbrian historian presents the Anglo-Saxon invasion of the island in a religious, Christian framework since he ascribes a very important role to the Anglo-Saxon people in the history of the Church and Christianity: in fact, while the Britons are seen by Bede as intrinsically heretical, the Anglo-Saxons, once converted to the religion of Christ, establish a new religious centre in England. In turn England becomes a place from which missions are sent to convert the Germanic peoples of the continent. As for the heretical tendencies of the Celtic people, in ch. I, 8, Bede falls back on the accusation of Arianism already leveled against the Britons by Gildas, one of Bede's sources, and also adds the accusation of Pelagianism. [12] The Anglo-Saxon invasion is seen by Bede as a step in the development of Christianity in Britain; in his view, the invasion was appointed by divine providence. He makes recourse to biblical rhetoric by means of allusions to Britain as the Promised Land and to the Anglo-Saxons as the Chosen People [13] destined for great achievements in the Christian Church. Bede understands the Anglo-Saxon conquest as willed by God and inscribed in a providential view of history. [14]

The reason for such a handling of the issue of the origins of Anglo-Saxon England lies in the author's Christian and missionary outlook. Bede's mind was dominated by Christian religion and also by the love of his people, the Anglo-Saxons, so that the *Historia Ecclesiastica* is a panegyric of the Anglo-Saxon church. [15] Bede in fact wrote in a missionary period when Northumbria itself was not yet fully converted; in addition, the Anglo-Saxon church, established by Augustine (who had been sent by pope Gregory the Great at the end of the sixth century) was closely connected to Rome and had to fight against the influence of

11 *Bede. A History of the English Church and People*, trans. and with an Introduction by Leo Sherley-Price; revised by R. E. Latham (Harmondsworth: Penguin Books, 1968), pp. 55–58.

12 Alhedys Plassmann, *Origo Gentis. Identitäts- und Legitimitätsstiftung in früh- und hochmittelalterlichen Herkunftserzählungen* (Berlin: Akademie Verlag, 2006), p. 63. Moreover, the Celtic church had a different way of calculating the date of the movable feast of Easter; according to Bede, the Celtic refusal of the Roman method of calculating Easter represented a menace to the universal religious supremacy of the Church of Rome and this is one of the main reasons for the accusation of heresy directed against the Celtic people of Britain.

13 Fabienne L. Michelet, *Creation, Migration, and Conquest. Imaginary Geography and Sense of Space in Old English Literature* (Oxford: Oxford University Press, 2006), p. 238.

14 Ibid., p. 257.

15 Ibid., p. 243.

the Irish missionaries in the north of England. [16] Bede was committed to the Roman cause in the dispute between the highly developed and centralized system of the Church of Rome and the Celtic monastic system. [17]

The *Historia Ecclesiastica* may also be seen as an attempt to legitimize the Anglo-Saxon domination over the original inhabitants, the Britons, who were politically and socially subjected to the Germanic invaders from the Continent, called collectively *gens Anglorum* by Bede. [18]

IV. The *Anglo-Saxon Chronicle*: the Anglo-Saxon Invasion as a Military Epic

Chronicles are the simplest form of history; they narrate the events in chronological order, year by year. [19] This annalist form derives from the practice of entering short historical notes on the so-called Easter Tables which were kept in every monastic centre and were drawn up for the computation of time. This was done to enable the clergy to ascertain the date on which the movable feast of Easter was to be celebrated. In the series of chronicles known collectively as the *Anglo-Saxon Chronicle*, we have a set of annals, written in Old English, which relate the major events in the history of Anglo-Saxon England.

The earliest entries begin with the creation of the world but are sporadic. The *Anglo-Saxon Chronicle* proper begins with the ninth century entries leading up to the Danish invasions of England and to the victorious stand of King Alfred of Wessex against these invaders. [20] It is probable that the *Anglo-Saxon Chronicle* originated towards the end of the ninth century in Wessex, as a reflection of both the 'revival of learning' and the resurgence of Anglo-Saxon national awareness during the reign of King Alfred. [21] There are seven extant manuscripts, produced in different monastic centres, falling into four distinct editions but growing out

16 See Henry Mayr-Harting, *The Coming of Christianity to Anglo-Saxon England* (London: Batsford, 1972).

17 Gransden, p. 18; Sherley-Price, p. 22.

18 See Nicholas J. Higham, *An English Empire. Bede, the Britons and the Early Anglo-Saxon Kings* (Manchester – New York: Manchester University Press, 1995). As for the name *Angli* used by Bede, see Patrick Wormald, 'Bede, the Bretwaldas, and the Origins of the *Gens Anglorum*', in *Ideal and Reality in Frankish and Anglo-Saxon Society: Studies presented to J. M. Wallace-Hadrill*, ed. by Patrick Wormald (Oxford: Blackwell, 1984), pp. 99–129.

19 Charles Plummer and James Earle, *Two of the Saxon Chronicles Parallel* (Oxford: Oxford University Press, 1929), p. XXIII.

20 Donald C. Scragg, 'Secular Prose', in *A Companion to Anglo-Saxon Literature*, ed. by Phillip Pulsiano and Elaine Treharne (Oxford: Blackwell, 2001), pp. 268–80 (p. 276).

21 *The Anglo-Saxon Chronicle*, trans. and ed. by Michael J. Swanton (New York: Routledge, 1998), p. XVIII.

of a common stock. [22] The earliest annals of the *Anglo-Saxon Chronicle* were written in Wessex, during the reign of King Alfred, probably at the command of the King himself. The Alfredian *Chronicle* covered the period from the creation of the world to 891, having as its main theme the history of Wessex; this is continued later, in sections of various length and content, in the different manuscripts. [23] In the case of the latter manuscripts, the events recounted extend to the second half of the twelfth century.

IV.I. The Entries Concerning the Anglo-Saxon Invasion

The Anglo-Saxon conquest of Britain is related in the first section of the *Chronicle*. The entries offer a detailed account of the military confrontations between the Germanic invaders and the original inhabitants, the Britons. All the entries concerning the arrival of the Anglo-Saxons are listed below:

449. In this year Mauricius and Valentinian obtained the kingdom [...]. In their days Hengest and Horsa, invited by Vortigern, king of the Britons, came to Britain at a place which is called Ypswine fleot at first to help the Britons, but later they fought against them.

455. In this year Hengest and Horsa fought against king Vortigern at a place which is called Agælesþrep, and his brother Horsa was slain. And after that Hengest succeeded to the kingdom and Æsc, his son.

457. In this year Hengest and Æsc fought against the Britons at a place which is called Crecganford and there slew four thousand men; and the Britons then forsook Kent and fled to London in great terror.

465. In this year Hengest and Æsc fought against the Welsh near Wippedesfleot and there slew twelve Welsh nobles; and one of their thanes, whose name was Wipped, was slain there.

473. In this year Hengest and Æsc fought against the Welsh and captured innumerable spoils, and the Welsh fled from the English like fire.

477. In this year Ælle came to Britain and his three sons Cymen, Wlencing, and Cissa with three ships at the place which is called Cymenesora, and there slew many Welsh and drove some to flight into the wood which is called Andredesleag.

485. In this year Ælle fought against the Welsh near the bank of Mearcrædesburna.

22 Plummer and Earle, p. XXIII.
23 Gransden, p. 32.

491. In this year Ælle and Cissa besieged Andredescester, and slew all the inhabitants; there was not even one Briton left there.

495. In this year two princes, Cerdic and Cynric his son, came to Britain with five ships at the place which is called Cerdicesora, and the same day they fought against the Welsh.

501. In this year Port and his two sons, Bieda and Mægla, came with two ships to Britain in a place which is called Portesmuþa, and slew a young Briton, a very noble man.

508. In this year Cerdic and Cynric slew a Welsh king, whose name was Natanleod, and five thousand men with him. The district was afterwards called Natanleag as far as Cerdicesford.

514. In this year the West Saxons, Stuf and Wihtgar, came to Britain with three ships at a place which is called Cerdicesora, and fought against the Britons and put them to flight.

519. In this year Cerdic and Cynric obtained the kingdom of the West Saxons; and the same year they fought against the Britons at a place now called Cerdicesford. And from that day on the princes of the West Saxons have reigned.

527. In this year Cerdic and Cynric fought against the Britons at a place which is called Cerdicesleag.

530. In this year Cerdic and Cynric obtained possession of the isle of Wight and slew a few men at Wihtgaræsburh.

552. In this year Cynric fought against the Britons at a place called Searoburh, and put the Britons to flight.

556. In this year Cynric and Cealwin fought against the Britons at Beranburh.

571. In this year Cuthwulf fought against the Britons at Bedcanford and captured four villages, Limbury, Aylesbury, Benson, and Eynsham.

577. In this year Cuthwine and Cealwin fought against the Britons and slew three kings, Coinmail, Condidan, and Farinmail, at the place which is called Dyrham; and they captured three cities, Gloucester, Cirencester, and Bath.

584. In this year Cealwin and Cutha fought against the Britons at the place which is called Feþanesleag, and Cutha was slain; and Cealwin captured many villages and countless booty, and departed in anger to his own [territories].

597. In this year Ceolwulf began to reign in Wessex, and ever fought and made war either against the Angles, or against the Welsh, or against the Picts, or against the Scots. [24]

IV.II. The Epic-heroic Elements in the *Anglo-Saxon Chronicle*

The entries which mention the arrival of the Anglo-Saxons indicate that the Germanic invaders fought a series of victorious battles for the possession of the land. The names of the chieftains, their family ties, the number of the ships each conqueror had, the names of the places where they landed and of the places where they fought are all recorded. In some cases, even the names of the enemies they slew are indicated. This points to oral poetry; the founding heroes (and their enemies as well) had names which were preserved (or elaborated) by oral tradition into the Christian period. While Bede depicted the violence of war and its consequences on the civilians, in the *Chronicle* war is only the confrontation of opposing armies on the battlefield, involving warriors and chieftains.

In the lexis of these entries, the verbs and the place-names are of particular frequency and importance. The invaders give a name to the places where they landed and fought which usually contains a reference to some of the Germanic leaders, such as *Cerdicesford* (so called after the chieftain Cerdic), *Portesmuþa* (after the chieftain called Port), [25] etc. This is one of the many cases of folk-etymology of place-names in the *Chronicle*; here Portsmouth takes its name from an Anglo-Saxon hero, and not from Latin *portus*, as it actually does. Folk-etymologies testify that the annals really reflect legends explaining the place-names. [26] To give a name to the places is another form, along with the military conquest, of taking possession of them; as Paul Zumthor put it: 'Nommer un lieu, c'est en prendre possession'. [27] Very common verbs in these entries are *feohtan* ('to fight, make war, combat'), *gefeohtan* ('to fight, to obtain by fighting'), *ofslean* ('to kill by a blow, to slay'), *geniman* ('to take, obtain', with reference to the land seizure), *ymb sittan* ('to besiege'), *gefleman* ('to cause to flee, to rout') and there is also the expression *bedrifan on fleame* ('to put in flight'). The grammatical subject of these verbs are always the Anglo-Saxons, while the only actions on the part of the Britons are expressed by the verbs *forletan* ('to

24 The translations of the passages from the *Anglo-Saxon Chronicle* are taken from George N. Garmonsway, *The Anglo-Saxon Chronicle* (London: J. M. Dent & Sons Ltd, 1972), pp. 12–15.

25 This is the town now called Portsmouth.

26 Patrick Sims-Williams, 'The Settlement of England in Bede and the *Chronicle*', *Anglo-Saxon England*, 12 (1983), 1–41 (p. 29).

27 Paul Zumthor, *La Mesure du Monde: Représentation de l'Espace au Moyen Age* (Paris: Editions du Seuil, 1993), p. 54.

leave') and *fleon* ('to flee'). In the diction of the *Anglo-Saxon Chronicle*, the Germanic invaders fight, conquer regions and towns, slay and rout the enemy, whilst the Britons can only take to flight. When, very rarely, an Anglo-Saxon warrior is reported to have died in battle, a passive construction (*Wipped wearþ ofslægen*, y. 465) or an impersonal one (*Horsan mon ofslog*, y. 465; *Cuþan mon ofslog*, y. 584) is used, without overtly expressing who the perpetrators (the Britons) were. Thus, words and grammar in the *Chronicle* are used ideologically, as a means to deny any active role on the part of the Celtic people.

In depicting the invasion, the *Chronicle* makes use of heroic diction, with particular attention to the glorification of the military prowess of the invading Anglo-Saxons and their martial exploits. [28] The entries concerning the battles fought against the Britons were probably made with the intention of supporting the established order by emphasizing its antiquity and the valour and success of the Anglo-Saxon rulers from the very start of their domination. [29] It is worth noting that Bede recounted a defeat of the Anglo-Saxons, that of Badon Hill (ch. 16), whereas in the *Chronicle* they are always the winners.

The *Anglo-Saxon Chronicle* was begun, as already said, at the command of King Alfred; in the ninth century a chronicle (now not extant) was drawn up in Wessex from numerous sources. The *Anglo-Saxon Chronicle* is in fact essentially a history of Wessex, with a touch of propaganda in favour of Alfred's dynasty. [30] The Anglo-Norman chronicler Gaimar wrote in the twelfth century that 'King Alfred had caused a book to be written in English about events and about laws and about battles in the land and about kings who made war'. [31]

It is noteworthy that the *Anglo-Saxon Chronicle* was produced during the reign and at the command of King Alfred, the most distinguished king of the house of Wessex. He had to face the Viking raids on England, which started at the end of the eighth century and had led to the Scandinavian occupation of the north of the country in the ninth century; in the year 886, with the treaty of Wedmore and after a lengthy military confrontation, Alfred succeeded in securing from the Scandinavian invaders the southern part of England. Thus, the *Anglo-Saxon Chronicle* was compiled at a time when the northern Anglo-Saxon kingdoms had succumbed to the Danes while King Alfred of Wessex had preserved his nation by successfully fighting against them. The historical situation of Alfred's time explains the patriotic fervour which runs through the *Anglo-Saxon Chronicle* from the very start – that is, from the account of the Anglo-Saxon invasion. The section of the *Chronicle* dealing with the battles for the land

28 Michelet, p. 238.
29 Gransden, p. 30.
30 Ibid., p. 34.
31 Swanton, p. XVIII.

which are at the origin of Anglo-Saxon England can be seen as an introduction to the history of Alfred's wars against the Danes, [32] since the conflict opposing Wessex and its king to the Scandinavian invaders is 'the very essence of the *Chronicle*'. [33] At a moment of a national crisis, the stress on Anglo-British fighting served a dual purpose: it legitimized and strengthened the power of the royal house of Wessex, which was emerging as the unifying authority for those areas in England not under Viking control, and it consolidated national awareness by way of memories of heroic times.

IV.II.I. The Entry for the Year 937: The Battle of Brunanburh

The memory of the wars against the Britons lingers in the *Chronicle* until post-Alfredian times, in the poem *The Battle of Brunanburh*. The continuation of the Alfredian *Chronicle* contains, in the annals from 925 to 975, a series of poems which celebrate events connected with Anglo-Saxon kings, such as a coronation or a victory. [34] *The Battle of Brunanburh* is one such poem. It celebrates the Anglo-Saxon victory at Brunanburh in the year 937. Led by king Ethelstan, the Anglo-Saxons defeated an army of Norsemen and Scots commanded by Anlaf and Constantine. The poem portrays the fight as a slaughter not seen in England since the Anglo-Saxons first came to Britain:

Ne wearð wæl mare
on þis eiglande
æfre gieta
folces gefylled
beforan þissum
sweordes ecgum,
þæs þe us secgað bec,
ealde uðwitan,
siþþan eastan hider
Engle and Seaxe
up becoman,
ofer brad brimu
Brytene sohtan,
wlance wigsmiþas,
Wealas ofercoman,
eorlas arhwate,
eard begeatan.
(ll. 65b-73) [35]

There has never been on this island a greater carnage in battle of people felled by the sword's edges before this one, as the books tell us, and the old learned men, since the Angles and the Saxons arrived here from the east, sought out Britain across the vast sea, the bold smiths of war, overcame the Welsh, the glorious and noble warriors conquered the land.

Thus, the poem reminds its audience of the Anglo-Saxons' battles for the land recorded in the earliest entries.

32 Michelet, p. 260.
33 Alfred P. Smyth, *King Alfred the Great* (Oxford: Oxford University Press, 1995), p. 478.
34 Gransden, p. 40.
35 Alistair Campbell, ed., *The Battle of Brunanburh* (London: William Heinemann Ltd, 1938).

IV.II.II. The Preface of the *Parker Manuscript* of the *Anglo-Saxon Chronicle* and Alfred's Genealogy

In the *Chronicle* the epic overtone is also found in the genealogy of King Alfred and in the legendary aura of his ancestors – whose martial exploits are discussed in the same entries that relate the Anglo-Saxon invasion. The *Parker Chronicle* is the oldest surviving version of the seven extant variants of the *Anglo-Saxon Chronicle*. While the E, D, F manuscripts have a preface based on the first chapter of *Historia Ecclesiastica*, the preface of the *Parker Chronicle* is different in that it gives an account of Alfred's genealogy that links him to Cerdic and Cynric, the two Germanic chieftains who founded the kingdom of Wessex. The genealogy of Cerdic and Cynric is then traced back to Woden, the Germanic god. [36] Power needs an origin. Descent from gods and demi-gods served an ideological purpose: the kings acquired political legitimacy, since they derived power and social prestige from these ancestors. The Parker preface states that Cerdic and Cynric *uuærun þa ærestan cyningas þe West Seaxna lond on Wealum geodon* ('were the first kings who conquered the land of the West Saxons from the Welsh') and then it proceeds to say that *þa feng Ælfred […] to rice, ond þa was agan his ielde .xxiii. wintra ond ccc. ond xcvi. wintra þæs þe his cyn ærest Westseaxna lond on Wealum geodon* ('Alfred succeeded to the kingdom […] when he was 23 years old and 396 years after his people first conquered the land of the West Saxons from the Welsh'). As can be seen, in the *Anglo-Saxon Chronicle* the memory of the invasion is continuously played upon.

Alfred's genealogy is also contained in the Latin biography written by his contemporary, Asser. [37] Asser traces the king's paternal descent, which goes back to Alfred's Germanic ancestors, Cerdic and Cynric and then, beyond them, to Woden, as in the *Parker Chronicle*. However, Asser goes beyond Woden to Adam, the first man according to biblical tradition. In doing this, Asser places Alfred in a wider Christian framework. In addition, Asser also traces Alfred's maternal genealogy: his mother, Osburh, was the daughter of Oslac, and Oslac *Gothus erat natione, ortus enim erat de Gothis et Jutis* ('Oslac was a Goth by race, for he was descended from the Goths and the Jutes'). With this allusion, the biographer associates Alfred with a wider Germanic world and includes in Alfred's back-

36 Bede mentions Woden as the ancestor of the kings of Kent, as well as of many other Anglo-Saxon dynasties (*Historia Ecclesiastica*, I, 15). Woden is the ancestor of the royal families in the Anglian genealogies, dating from the second half of the seventh century: see Kenneth Sisam, 'Anglo-Saxon Royal Genealogies', *Proceedings of the British Academy*, 39 (1953), 297–332, and David N. Dumville, 'The Anglian Collection of Royal Genealogies and Regnal Lists', *Anglo-Saxon England*, 5 (1976), 23–50; Henry Moisl, 'Anglo-Saxon Royal Genealogies and Germanic Oral Tradition', *Journal of Medieval History*, 7 (1981), 215–48.

37 Asser, *Life of King Alfred*, ed. by W. H. Stevenson (Oxford: Clarendon Press, 1904; repr. 1959), esp. chapters 1 and 2.

ground the continental homeland of the Germanic tribes. [38] Thus, the issue of origins (the account of the arrival of the Anglo-Saxons and the royal genealogies) plays a key role in the legitimation of power, in both the *Chronicle* and Asser.

Another element in the *Anglo-Saxon Chronicle* which contributes to giving an epic-heroic role to the Germanic newcomers, and, as a consequence, to their descendants on the throne of Wessex, is the issue of their territorial origins. The *Parker manuscript*'s preface does not specify Cerdic and Cynric's place of origin A later hand, dating from the post-Norman conquest era, has inserted an interpolation, based on Bede's account, which locates the homeland of the two leaders on the continent. However, the original compiler of the preface did not mention their homeland, in this way exploiting the traditional motif of the mysterious ruler, of unknown origins, who comes from beyond the sea. This legendary motif is also found in the Anglo-Saxon epic-heroic poem *Beowulf*, in the story of Scyld Scefing, the mysterious ruler who came to the Danes on a boat as a child and then established kingship in Denmark. [39]

IV.II.III. The Sources of the *Anglo-Saxon Chronicle*

The sources for the entries in the *Anglo-Saxon Chronicle* dealing with the invasion and the battles against the Britons are obscure. The *Chronicle* was a composite document, drawing on several sources. Some early entries were taken from the chronological summary appended by Bede to the *Historia Ecclesiastica*. [40] There was probably a set of West Saxon annals starting with the narration of the Anglo-Saxon invasion, which were used in the Alfredian *Chronicle* when Alfred had the idea of creating a national chronicle as opposed to merely local annals. [41] The early section of the *Chronicle* that employs epic-heroic and legendary motifs may have incorporated oral traditions, which had been partly in the form of epic poems. [42] The use of oral poetry is also suggested by the use of alliteration, as in the names of Hengest and Horsa, Cerdic and Cynric, and Cuþwine and Cealwine. Palatalization had not yet taken place at the time of the invasion, so that the initial 'c' in *Cerdic* and *Cealwine* was still an unvoiced velar plosive in the sixth century, as in *Cynric* and *Cuthwine*. It is therefore highly

38 See *Alfred the Great: Asser's Life of Alfred and Other Contemporary Sources*, trans. and ed. by Simon Keynes and Michael Lapidge (London: Penguin, 1983), pp. 229–30; Janet L. Nelson, 'Reconstructing a Royal Family: Reflections on Alfred, from Asser', in *People and Places in Northern Europe, 500–1600*, ed. by Ian Wood and Niels Lund (Woodbridge: Boydell, 1991), pp. 47–66.

39 *Beowulf and the Fight at Finnsburg*, ed. by Frederik Klaeber (Lexington, Massachusetts: D. C. Heath & Company, 1950), ll. 4–52.

40 Plummer and Earle, p. CXIII.

41 Swanton, p. XIX.

42 Gransden, p. 37.

probable that some narrative material in the entries relating the invasion had derived from oral epic-heroic traditions of the Anglo-Saxons.

V. The Anglo-Saxon Origins: Two Different Paradigms

The accounts of the origins define identity and legitimize a social and political order; they also transmit the life rules of a people. [43] These accounts can be embedded in different historical works, as testified by the two Anglo-Saxon sources; as a consequence, the interpretations of origins, and the values they endorse, are different because they are affected by different contexts. Origins are used in different ways, according to different points of view. In both works the Anglo-Saxon invasion sets the tone of subsequent Anglo-Saxon history and society, which are religiously characterized by Bede, while in the *Chronicle* they are depicted as epic-heroic; the two works offer different paradigms of origins and, as a consequence, different views of Anglo-Saxon identity.

It can be said that both the *Historia Ecclesiastica Gentis Anglorum* and the *Anglo-Saxon Chronicle* aim to legitimize Anglo-Saxon rule in Britain, which dispossessed the Britons of their land; however, the grounds for such a legitimation are different in the two works.

The work of Bede presents the religious history of his people. He lived in 'the golden age of Northumbria', a time of relative peace as compared to the times of Alfred. Bede's account of origins offers a paradigm focusing on a Christian people chosen by God, a paradigm he then develops in the following history of the Anglo-Saxons in the *Historia Ecclesiastica*. The identity of the Anglo-Saxons defined by the Christian mission and by the close link to the Church of Rome is an unusual one among the *origins* [44] of the mediaeval Germanic peoples. The *Historia Ecclesiastica Gentis Anglorum* presents the Britons as heretical, so that the *gens Anglorum* who defeats them and causes the orthodoxy to prevail in Britain is portrayed as the Chosen People, owing to its religious superiority. Bede makes recourse to biblical rhetoric, focusing especially on allusions to the Promised Land and on the identification of the Anglo-Saxons as the Chosen People. Being primarily preoccupied with religious issues of the Anglo-Saxon church, in his *Ecclesiastical History* he highlights the Anglo-Saxons' religious destiny and religious achievements.

On the other hand, the *Anglo-Saxon Chronicle* makes use of a military rhetoric with particular attention paid to the glorification of the military conquest; it

43 See Karl Hauck, 'Lebensnormen und Kultmythen in germanischen Stammes- und Herrschergenealogien', *Saeculum*, 6 (1955), 186–223.
44 Plassmann, p. 83.

values heroic ethos and offers a legendary view of the beginnings of Anglo-Saxon England. This text gives a detailed account of the battles fought and won by the Germanic invaders against the Britons; in it, the right to land is based on military supremacy.

The *Anglo-Saxon Chronicle* does not inscribe the invasion in a divine scheme, it had no edificatory purpose in a religious sense; it was primarily a secular record, whereas Bede's historical work was a by-product of his religious pre-occupation. In the *Anglo-Saxon Chronicle* the memory of the Germanic tribes' invasion helped to generate a sense of community among their descendants; Bede's idea of a united England was instead dominated by his belief in the English church. His historical work is a panegyric of the Anglo-Saxon church, while the *Chronicle* is a panegyric of Anglo-Saxon military kingship.

Marianne Børch

The Semantics of Originality as Indicator of the Cultural Paradigm Shift from Medieval to Modern

When St. John wrote 'In principium erat verbum' about the Word of God begetting Christ, he was continuing in the footsteps of the Old Testament mythographer, who has God speak Creation into being. Today we say the same: in the beginning is the word. Even the most rabid atheist will acknowledge that whatever we experience or know is mediated by language and thus to some extent created by the word.

The referents for the new word and the old are, of course, fundamentally different. Between the two eras of the word there has been a short, but momentously important period of 350 years when verbal language has been viewed as an obstacle to true knowledge at worst, and at best as ideally transparent, leaving no imprint upon the known, or as a refined game of fiction or poetry, whose value it is impossible to define, and in any case in need of defence. The period is called the Enlightenment, and the abandonment of the linguistic universe coincided with the rise of the empirical sciences. Meanwhile, the so-called arts or humanities gradually lost legitimacy to such an extent that today they are in deep crisis. We hear that:

the exact and unquestionable truths of science do not encompass the humanness of the human subject, which demands a more than empiricist-rationalist approach;

innovative scientific thought may be achieved only through leaps of imagination which, although based on hard data, can hardly be distinguished from those of poetic creativity;

sciences in isolation from the context they are part of are barren and thus have to develop holistic strategies;

play, improvisation, and other activities undirected at a specific serious goal may generate ideas of a scientific nature, and may valuably be undertaken by those who want to prevent their thought from becoming stuck in futile hermeneutic circles;

the words and images that are necessarily employed in communicating what we know inevitably determine what we can know and the questions we ask.

The latter point about imagery as a conceptual vehicle can be elucidated by noting how different it is to describe the attack of white blood cells upon bacteria as that of a valiant rescue team, or as a merciless, cruel onslaught. We do not often empathise with bacteria, and yet it is possible. One successful case is Norwegian Thorbjørn Egner's *Karius and Bactus*, a book which stimulates children to brush their teeth. [1] Despite the book's ultimately anti-bacterial stance, it makes us identify wholly with Karius and Bactus, who are building nice houses in Jacob's teeth and enjoy a good laugh at the groans their drilling elicits from the boy whose mouth they are raiding. In the end, however, Jacob goes to the dentist, and poor Karius and Bactus are flushed out after their fine houses, Jacob's cavities, have been filled. The unwonted solidarity with the bacteria here is disconcerting, but we also get to think in a way that might bring about new insights – and some of these sudden impulses or leaps might be so cognitively disruptive as to open new, arguably creative, avenues of thought. [2] To discover, to invent, and to create are related activities, and because they all depend not just upon finding something new, but importantly upon breaking the mould of the old, an unusual image or a narrative inverting the traditional perspective may spark the new idea.

Today, we seem to have almost stopped using one particular image that was once the central cultural trope for Western thought: that of the book. [3] We increasingly talk of culture in terms of texts and textuality, not of culture as a book, as we used to do. Indeed, different imagery identifies different world pictures.

Christianity and its Jewish parent, as well as its cousin Islam, are religions of the book, their laws and mythologies being inscribed in sacred books. According to the Christian creed, God revealed his narratives, injunctions, and truths to Moses and others who wrote them down in God's first book, the Bible. God also created a second book, the world itself.

The Middle Ages posed that 'science', or human knowledge, came from reading the books of God – but specifically in such a way as to correlate their respective kinds of evidence: the different books had, in other words, to confirm one another. People struggled to read the Bible as a single and seamless whole

1 Thorbjørn Egner, *Karius og Baktus* (Oslo: Gyldendal, 2001; first edition 1949). For the English version see <http:*//www.youtube.com/watch?v=VOJo73IGpfE*> [accessed 3 September 2009].

2 For an author who uses disease, otherwise consistently conceptualised as negative, in order to explore the overwhelming feeling of love, read Jeanette Winterson, *Written on the Body* (London: Vintage, 1992). For a discussion of Winterson's cancer metaphorics see Marianne Børch, 'Love's Ontology and the Problem of Cliché', in *Sponsored by Demons: The Art of Jeanette Winterson*, ed. by Helene Bengtson, Marianne Børch, Cindie Maagaard (Odense: Scholars' Press, 1999), pp. 41–54.

3 See e.g. Jesse Gellrich, *The Idea of the Book in the Middle Ages* (Ithaca: Cornell, 1985).

and saw the physical world as a single text where everything fitted in with everything else, without contradiction; it corresponded with Biblical truth and offered up empirical analogues of it. The duty of theologians – who were the scientists in so far as we accept that word in this context – was to wrest things into a single coherent unity. For this purpose they had words at their service.

Of all created things and beings, only man has language, necessary for the communication of reason, and thus the correlative gift of man's reason, which separates him from beast below him, and angel above him. [4] As a creature communicating his reason by the word, man is the microcosmic carrier of the very Word by which God created the world. This conviction led to another proud idea: the human investigator should properly assemble every possible kind of knowledge, and the result is an analogue of the world itself, or of God's Bible, a man-made and imperfect, but still analogously worthy and valid, third book. Hence the attempts at books that included everything – typically called a *summa*, *cosmographica*, *speculum mundi*, and, especially later in the Renaissance, an *anatomy*. Such works came from universities, monasteries and other religious communities whose studies encompassed everything; significantly, the Bachelor of Arts *trivium* consisted of the *linguistic* disciplines of grammar, logic and rhetoric, the premises of all understanding and creation to be acquired prior to the Master *quadrivium* disciplines of astronomy, arithmetic, geometry, and music. The *artes liberales* set men free to understand God's cosmos in the proper way.

The medieval context stimulates three questions:

What were the implications for medieval 'science' of the not only linguistic, but book-like nature of things?

What may be originally discovered or invented in such a world?

How did the perception of all things as part of a single harmonious and God-given entity determine the way things were represented?

These questions can be answered by focusing on the agonies of late-medieval culture: the gradual detachment from old paradigms and the search for new ones was no less frightening than liberating.

To discover true knowledge, instead of beginning with an open mind, medieval scholars had to start from a clear expectation about what they would find, and even had to find. If the Bible says that the hare is an animal that chews the cud – like a cow – then we have to find out that a hare chews the cud. We seek out evidence that may corroborate the Biblical truth, and anything that might contradict it, either in the Bible itself or in external evidence, has to be subjected to a harmonising approach. In a famous passage, Boccaccio insists that we

4 A locus classicus for this idea is Dante's *De Vulgari Eloquentia*, ed. by Pier Giorgio Ricci (Firenze: Le Monnier, 1968), I, II, 1 – 2.

should go on reading until we find the desired proper meaning even in arcane or intransigent material.

> You must read, you must persevere, you must sit up nights, you must inquire, and exert the utmost power of your mind. If one way does not lead to the desired meaning, take another; until if your strength holds out, you will find that clear which at first looked dark. [5]

Boccaccio is himself the child of St. Augustine, who believed that a text which refused to cohere with Christian truth should be read allegorically, a procedure that was to save the pagan heritage for Christianity. [6] A story from the *Gesta Romanorum* of a man whose wife hangs herself in a tree might, for instance, be construed as the way a Christian crucifies the weakness of his flesh (his wife) by fastening it to the tree of Christ's cross. [7]

Augustine's approach also explains the late-medieval astronomers' attempts to 'save face', an expression that is today proverbial for all kinds of manipulation that manage both to promote the necessary truth and make this look consistent with what we experience, but was then an attempt to discover solutions that accounted for astronomical observations within the necessary doctrinal premises. We tend to scorn the authorities who resisted a heliocentric world view and stopped the mouth of scientists like Galileo, but we should try to understand them as they struggled to salvage truth as a human-centred conception. The roundness and finite size of the geocentric cosmos made human sense, while Copernicus undermined it. Moreover, the challenge of visible enigmas to human understanding gave rise to elegant and highly sophisticated mathematical computations and representations of the world, as we know from early models of the universe. These honed and perfected human sciences in multiple ways, while working to bridge two entirely incompatible truths. In this process there was, of course, no distinction between science and faith, nothing to separate the strictly investigative process from the inventive, creative power, which the Middle Ages viewed as supra-rational or visionary. Medieval scientists or liberal artists –

5 *Boccaccio on Poetry: Being the Preface and the Fourteenth and Fifteenth Books of Boccaccio's 'Genealogia Deorum Gentilium'*, trans. and ed. by Ch. W. Osgood (Indianapolis and New York: Liberal Arts Press, 1956), p. 62.

6 The Biblical tribe of Israel took away with them Egyptian gold on their Exodus and Christians were allowed to preserve and perpetuate Classical texts and culture, as long as they used the old treasures for new and legitimate purposes. For an exposition and discussion of this principle of medieval exegesis, see D. W. Robertson, *A Preface to Chaucer* (Princeton: Princeton University Press, 1962), chapter 2.

7 This 'moralisation' demonstrates the way proper reading allowed for total disregard of textual or narrative logic in the vehicle story. For a discussion of the example given, see Helen Cooper, *The Structure of the Canterbury Tales* (London: Duckworth, 1983), pp. 10–11.

working creatively with analogy, the master Creator's own activity – were groping for the leap or spark by which two incompatible data, both incontrovertible, suddenly stand forth as continuous and in mutual harmony.

When they made this leap, they *invented*: in the Middle Ages to invent meant to find what is already there, an essence waiting to be found out. The word *discover* contains the same etymological referent to the already there: to discover, take the lid off, and find what is really, truly there. Thus, theologians invented the beautiful truth always waiting to be found in the data. Although the result of one's quest was already given, the cases cited above – of detecting Christian truth inside Classical pagan legend or a geocentric world in heliocentric appearances – show that invention was by no means a stale and dull faculty, but might require a great deal of creative ingenuity.

Until a few decades ago art historians held that the Middle Ages had no aesthetic theory.[8] However, this view rested upon false assumptions about creativity, for invention as 'finding what is there' undoubtedly requires creativity on a par with modern expectations about free and spontaneous expression. Furthermore, the scholars' failure to uncover medieval theories of art was due to the fact that they searched in the wrong places. Instead of seeking aesthetic criteria 'inside' works of art, we should realise that the medieval artefact is prior to human activity, i.e. that the medieval world is itself conceptualised as aesthetically perfect. Creation is God's art and so has to be understood through science in-venting, or dis-covering, its patterns and truths. This creative process was performed not by the imagination, which is where moderns look for direction on aesthetic matters, but by memory. While processing what we learn or perceive so that knowledge becomes personal, memories are creative in the medieval sense of inventing what is already there.[9] The artist, then, would use this creative memory to explore and reproduce already available truths in representations considered artistic and creative, but 'original' only in so far as they managed to recreate or invent in the present the 'original' and ideal instantiation of a thing. In other words, the closer they got to the single, coherent, and original conception, the finer their invention.

This aesthetics involves a preference for the tropes of the simile or comparison which, by pointing to sameness and definable likeness, identify an already existing ontological bond between tenor and vehicle, and ultimately between both these and God's cosmos, whole, comprehensive, and perfect. The cathe-

8 The traditional view that the Middle Ages 'was lacking any overt aesthetic standard' is based precisely on the fact that art was viewed as ontologically prior to human modes of representation: the medieval 'aesthetics was an appendage of a theological system which was external and alien to aesthetic considerations': Harold Osborne, *Aesthetics and Art History* (New York: Dutton, 1970), pp. 132–34.

9 See Mary Carruthers, *The Book of Memory* (Cambridge: Cambridge University Press, 1992).

drals were built as microcosmic analogues of the entire world; hence the necessary representation in the cathedral of all possible things in the world, grotesque and lewd as well as sacred and pious. Music was theoretically composed after Pythagorean mathematical principles so as to reproduce the music of the spheres. It sounds good in so far as we are able to reconstruct it, but it was principally created not for human delight, but to tune the listener's moral instrument to the harmonies of the cosmic order; a matter not of aesthetic appeal, but of synchronicity in analogously created things, by which one level duplicates the other. Our vocabulary for peace and war still retains the echoes of this idea as we speak of concord and harmony versus discord; even today we think music might appease wild emotion, and that cows yield more milk if Mozart is played to them, and possibly less, if they are treated to heavy metal.

Allegory, which is recommended by St. Augustine and also forms Thomas Aquinas' epistemological base, builds upon analogy, not metaphor. [10] Although metaphor and analogy are often classed together as imagery or tropes, they are radically different, signalling entirely different conceptions of the world and cognitive avenues to its understanding. Their use is also differently distributed in historical time: analogy, simile or comparison goes with the medieval and renaissance world – and is, significantly, making a comeback today; whereas the heyday of the metaphor characterised the post-seventeenth-century world and is itself a symptom of the epistemological catastrophe by which science as true and exact knowledge parted company with human-centred disciplines.

An example will clarify the crucial difference between metaphor and simile: 'My home is my castle' is a metaphor, and the response to this self-evidently incorrect statement has to be subjective, irrational, and loosely exploratory, for there is no logical way to bridge the gap between vehicle and tenor. Indeed, the strong effect of metaphor builds upon the dissimilarity between constituents that are claimed to be identical.

The simile operates in a different manner: the simile 'My home is like a castle' has common reference to things we can see and whose features we may list in shared, literal, and rationally analysable terms. On consideration, a home might in fact somewhat resemble a castle – not in terms of physical features, but as a protective haven, which makes me feel safe. The simile, then, points to identifiable similarity and downgrades difference, since it does not invite attention to ways in which vehicle and tenor are dissimilar. Whereas metaphor is the irrational leap that struggles to assert the truth of the imagination against that of

10 For the theological/epistemological implications of medieval analogy, see Marcia Colish, *The Mirror of Language: A Study in the Medieval Theory of Knowledge* (New Haven: Yale University Press, 1968).

reason and observation, analogy or simile are quite compatible with an empirical and a commonsensically rational, culturally constructed epistemology.

Nevertheless, the Enlightenment downgraded analogy as a cognitive device. Attention was directed towards the concretely individual, particular, or specific as the basis for any understanding of the general; in other words, the inquirer progressed from individual phenomenon to general principle rather than from an overall order to things within that frame. Here was a point upon which, despite their different implementations, scientists and progressive artists agreed. Between 1769 and 1790, the neoclassical painter Sir Joshua Reynolds addressed to the Royal Academy a series of discourses in which he asserted the ancient principle that the defining mark of the true painter was stylised representation of the general in the unique. According to Reynolds, 'Invention, strictly speaking, is little more than a new combination of those images which have been previously gathered and deposited in the memory. Nothing can come of nothing' [11]. In the margins of his copy of Reynolds' *Discourses*, however, William Blake wrote: 'To generalize is to be an idiot. To particularize is the alone distinction of merit – General knowledges are those knowledges that idiots possess'. [12]

Why this dislike of analogy? If the simile is a literalist and logical trope, and if the world is analogically indicative of the truth, one might expect simile and analogy to serve the new empirical sciences well. However, in medieval way of thinking analogy was regarded as a closed quasi-linguistic system which predetermined meaning and what might legitimately be in-vented; it forestalled any attempt to dig freely into nature, because the answers to the inquiry were already given. Analogy not only defines, but also fixes and restricts. Mention of one element in the cosmos simultaneously implies all the other elements as well as the meaningfulness of the implied whole.

Even if empirical investigation was slow in establishing itself, realistic representation of external reality had little interest for artists. References to architecture and music have already been made, but they should be supplemented by some observations on painting and poetry.

A striking feature of medieval art is its preference for formulaic phraseology and iconography. Once more, the repetition and similitude which rule in this

11 Joshua Reynolds, 'Discourse II. Delivered to the Students of the Royal Academy, on the Distribution of the Prizes, December 11, 1769, by the President', in Sir Joshua Reynolds's *Discourses*, ed. with Notes and Historical and Biographical Introduction by Edward Gilpin Johnson, with Illustrations (Chicago: A. C. McClurg and Company, 1891), p. 66.

12 William Blake, 'Annotations to The Works of Joshua Reynolds', ed. by Edmond Malone, in *The Complete Poetry and Prose of William Blake*, ed. by David V. Erdman with a New Foreword and Commentary by Harold Bloom (Berkley and Los Angeles: University of California Press, 2008 [first publ. 1965]), p. 641.

aesthetics serve to approach as closely as possible the art or image spoken by God 'in the beginning'. Whether you look at the Garden of Eden or classical myth, the world had fallen or declined, and art's duty was to highlight the initial plan that had been obscured by sin, original or personal.

This explains why all portraits, verbal or pictorial, are highly stylised and downgrade what we might call individually differentiated traits. For instance, by medieval convention women would ideally be portrayed according to a so-called 'descending catalogue', starting from the top and going down through hair, forehead, nose, hue or skin colour, mouth, chin, neck, breasts, waist, hips, etc. Moreover, each of these features has stock epithets to go with it: foreheads are high, mouths small, round and red as rose, skin fair as lily or snow, eyes 'grey as goose' or 'grey as glass', breasts high and round. If one asks how such portraits relate to the sitters, the answer is that they bring out their ideal femininity. Today we have a low opinion of clichés, but for the medieval person a cliché is again ontologically prior to, and more perfect than, any individual, although the idea does not exclude individuality. The question raised by a famous painting in which Jan Eyck represents Agnes Sorel as the Virgin Mary is whether an identifiable royal mistress of the painter's own time and the Virgin Mary are not incompatible. They are not: the artist spans the apparent gap as he invents the original perfect woman – Mary – immanent in the latter-day, temporally finite and individual woman Agnes. He invents Mary's original image in its new manifestation, the original Agnes.

The word 'woman' is habitually used in an absolute sense denoting a perfect conception, rather than describing ordinary and variable human lessness or imperfection. Rather than being termed a 'fine horse' or a 'fast horse' or a 'beautiful horse', a first-class horse would be characterised as a 'horsely horse', because the horse in question coincides very nearly with the idea of a horse. A powerful or strong man is a 'manly man', a good monk a 'monkly monk', and indeed medieval literature abounds in 'knightly knights' and 'womanly women'. All these terms are latent similes, abbreviated versions, as it were, of 'as horsely as a horse', or 'as manly as a man'. They depict an ideal world – but they also easily signal irony: for if you say a monk is a 'manly monk', you indicate a clash between his ideal type and his actuality. Even variation is severely restricted: if a knight is not called knightly, he is invariably 'noble'. Where modernity tends to pursue variety of expression, the word noble repeated *ad infinitum* identifies the quality which drives all knights and testifies to their ideal embodiment of God's nobility. [13]

13 For the homogenising effect of analogy, see Derek Brewer, 'Chaucer's Poetic Style', in *The Cambridge Chaucer Companion*, ed. by Piero Boitani and Jill Mann (Cambridge: Cambridge University Press, 1986), pp. 227–42; Jill Mann, 'Malory: Knightly Combat in Le Morte

If modern readers do not understand this usage, they are likely to read early texts as naïve, trite or plain stupid. More seriously, they fail to understand how the world looked back to earlier ages, exploring how the 'real' and its meaning shone forth in the empirical. Thus, in a saint's legend every element has to be accommodated within the prescribed category before it makes sense within the overall framework, and the result is odd to a modern perception. For instance, in Chaucer's retelling of the legend of Saint Cecilia from the *Canterbury Tales* (c. 1385–1400) her husband Valerian is singularly sympathetic; however, he *is* a pagan, so when the writer describes the saint's beneficial influence as a converter of heathens, Valerian is represented as a demonic 'fiers leon' (mad lion).[14] He is a meek lamb in action, but a mad lion with respect to his function in the analogically constructed cosmos; his function overshadows his sympathetic individual features, and rather than exploring complexities or tensions within an interesting character, the text restricts possible significance into its needed meaning. Such a procedure is illogical in a purely narrative optic, but is true to the story's epistemological assumptions. Totally dissimilar descriptions in different situations express an aesthetics which can perceived as flawed by modern readers, but merely follows a different logic.

As for Cecilia herself, she, too, is rather odd. Cold as ice in her scorn of worldliness, but nevertheless, we are told, 'round' and 'brennynge'. Fleshy and burning? Of course not; she is as round as the stellar sphere and as intense as the sphere of fire of which she is the analogue. Being perfect, Cecilia is the human embodiment of the principle also embodied in the corresponding sphere; woman and sphere are created by the same Logos and thus analogous in moral significance. Everything is itself and evokes everything else at the same time. The cosmic may be found, or invented, in the personal and vice versa. The verb Chaucer uses here is 'declare' (l. 119), which does not mean assert, but clarify; indeed, speaking about Cecilia in the right words means casting light upon her as a function of the cosmos.

However, towards the late Middle Ages culminating in the fourteenth century, artists begin to use words with new referents. Significantly, this is also the period which saw the great philosophical debate between the so-called Nominalists and Realists about the referents of words as either absolute and God-given, or arbitrary and socially constructed – a debate cleverly exploited by Umberto Eco in *The Name of the Rose* (1980). Again Chaucer provides an example, this time with his portrait of Alison in *The Miller's Tale.* Here the female protagonist is de-

d'Arthur', in *The New Pelican Guide to English Literature*, 8 vols, ed. by Boris Ford (Harmondsworth: Penguin, 1982), I. Medieval Literature. Chaucer and the Alliterative Tradition, pp. 332–35.

14 All quotations from Chaucer are from *The Riverside Chaucer*, ed. by Larry C. Brewer (London: Oxford University Press, 1988).

scribed as similar not to the spheres, the Virgin Mary, or a set of standard flowers with standard epithets ('red rose' or 'white lily'), but to domestic animals and plants (e. g. swallow, calf, primrose, apple), and newly minted London coins – all local, concrete and social referents. Alison lives in Oxford, a seat of academic scholarship with alluring and available clerks, but also a semi-rural town, in which each house is a self-supporting unit with farmyard animals, workshop and student tenants. The portrait of Alison is thoroughly sensuous: she is as soft as the wool of the wether, she sings as readily as the swallow, she moves as fast as a kid or colt, her mouth tastes and her body smells as good as mead or cider, and her looks are gorgeous – and highly unusual. The traditional woman, always gazed upon and represented as static and defined within the necessary network of correspondences, is replaced with a fast-moving, singing, and skittish little animal; on top of that, Alison is not only looked at, rather she herself looks back at the viewer with a 'likerous' (i. e. lecherous) eye. Finally, the observer voices his delight by saying that she is just as delightful to look at as a young pear tree in spring; the comparison, which refers to an observer's feel-good response rather than any precise likeness between a pear tree and a woman, operates in the manner of metaphor.

When the same author creates both the sensuously appetising Alison and Cecilia the round and burning sphere, the difference between the portraits cannot be explained only in terms of genre consciousness. Indeed, the description of Alison marks an assault upon the medieval world picture as such, for she cannot in any way be contained within the cosmic order: her portrait bypasses all the usual epithets with their paradigmatic referentiality and evokes a character understood in terms of her social and material context. Moreover, owing to her amoral, rather than immoral, drives, Alison blows up the binary taxonomy of Christian morality versus immorality. She alone emerges victorious at the end of this narrative, which is undoubtedly the most hilarious, but also certainly the naughtiest story composed in the English language. As Alison simply eludes the categories of morality, her escape from punishment becomes logical, for analogical thinking, which would call down damnation over her head, has been dismissed as irrelevant.

Such usage explodes the framework for categorising things strictly in precise relation to every other ingredient of the cosmic whole and paves the way for a new conception of knowledge by breaking with the necessary repertoire of fixed similes and correspondences. Artists whose inventive ventures involved recreating the original, the first and ideal, found that their inventions moved into areas that could be held neither within the binaries of Good versus Evil, moral versus immoral, true versus false, inventive versus innovative, nor within the prescribed qualities allotted to the individual creature in the scale of being. An animal-like woman is no longer automatically bestial or demonic, but blows up moral categories altogether. Art, the realm of narrative and simile which become problematical when they disrupt rather than confirm traditional analogy, is where a new possible idea of the world may be tested out before its findings can ever break through to become the basic premise for understanding the world; for something must be imagined, before it can even be sought, let alone found, in-vented, or dis-covered. Art is thus one of the places where the preconditions of science as we know it begin to emerge. [15] At the end of the Middle Ages every word had to tear loose from its predetermined place in the cosmic whole before it could signify freely things as such, or operate in a new relation of individual intention and curiosity. And the new, liberated word was thought to refer properly to things rather than to a linguistic network ultimately grounded in God's Logos.

A fundamentally new way of thinking does not develop all at once. Moreover, survivals of earlier practices, along with past evidence, provide us with indexes to truths that seem at variance with our own conceptions and so may be hard to detect and respect. If we acknowledge that today's interpretive models suit our predispositions, we also have to mitigate our scorn of earlier world views. We laugh because it took two hundred years for Copernicus' heliocentric model to be accepted, but again, we need to imagine why they thought as they did. Our own premises, as well as those of former times, are constantly overcome by the very actualities we study. At the same time, however, they show our creative will to invent order and meaning.

Geocentricity had to be salvaged because it was not only God's truth, but hermeneutical, human-centred truth; it was a false, but meaningful image. What came afterwards was an objective representation expressing empirical and scientifically measurable correctness, with meaninglessness gaping beyond. People could see that empiricists were right, according to the senses, but the demand for a commensurability between evidence and the necessary textual logic of God's book stalled the scientific breakthrough for at least two hundred years.

The change came in the seventeenth century, and people agonised over it. In *Paradise Lost* (1667) John Milton plays out the drama of the Fall in Ptolemy's humanly meaningful cosmos, but the archangel Raphael nevertheless suggests

15 For discussions of the late-medieval breakdown of analogy in literary texts – which also amounts to the invention of 'literature' as an independent avenue for cognitive inquiry, see Jesse Gellrich, *The Idea of the Book in the Middle Ages*, and Marianne Børch, 'Chaucer's Second Nun's Tale: Record of a Dying World', in *A Chaucer Yearbook 5*, ed. by Jean Jost and Michael Salda (Brewer: Cambridge, 1999), pp. 19–40; Id., 'Chaucer and the Cosmic Text: Rejecting Analogy', in *Text and Voice: The Rhetoric of Authority in the Middle Ages*, ed. by Marianne Børch (Odense: University Press of Southern Denmark, 2004), pp. 97–120.

to Adam that the universe is, in fact, heliocentric. [16] Earlier in that century, the poet John Donne stretches similes to the limit, as he squeezes the early sciences searching for analogies that will keep knowledge and desire together. For him the appeal lies in the simile's logical link between vehicle and tenor, reference and referent, not in the evocative fuzziness of metaphor. While Donne is trying to find ever more adventurous ways of getting things to cohere and speak one through the other, the referents become more and more strained, and when later critics read such poetry, they decided to call them 'metaphysical', although they were in fact scientific. Yearning to keep together the world outside and the world inside, Donne knew that he lived at the end of an age, when things fell apart and the centre could no longer hold. In 'An Anatomy of the World' (1611) – note the holistic image of the world as a body to be anatomised – he laments the loss of a single and harmonious cosmos and its replacement by a fragmented world where things float freely without direction or proper place, just things, unconnected by the meaningful, relation-creating, word.

And new philosophy calls all in doubt;
The element of fire is quite put out;
The sun is lost, and the earth, and no man's wit
Can well direct him where to look for it.
And freely men confess that this world's spent,
When in the planets, and the firmament
They seek so many new; they see that this
Is crumbled out again to his atomies.
'Tis all in pieces, all coherence gone;
All just supply, and all relation:
Prince, subject, father, son, are things forgot,
For every man alone, thinks he hath got
To be a phœnix, and that there can be
None of that kind, of which he is, but he.
This is the world's condition now [...]. [17]

Donne here ascribes the expanding universe of senseless atoms to the death of a single, but typologically perfect woman, Elizabeth Drury. She was the analogical entity whose presence guaranteed the existence of the others:

16 John Milton, *Paradise Lost*, ed. Gordon Tesky (Norton: London & New York, 2004), book VII, pp. 205–19.

17 John Donne, 'The Anatomy of the World', in *The Works of John Donne, D. D., Dean of Saint Paul's, 1621–1681. With A Memoir of His Life*, ed. by Henry Alford, 6 vols (London: John W. Parker, 1839), VI, pp. 491–92.

She that was best, and first original
Of all fair copies, [...]
She to whom this world must itself refer
As suburbs, or the microcosm of her,
She, she is dead [...]. [18]

This sign, or analogical index, being lost, the particles escape the force field of atoms in mutual tension and balancing attraction to explode outwards in every possible direction, a movement that, although a new certainty, is, in its lack of relation and form, deprived of human meaning.

The change means loss, but it also opens up knowledge which does not fit into the system. The insecurity of this position should not be underestimated: the Enlightenment motto of *Sapere aude* was spoken to people understandably nostalgic about metaphysical certainty and protective holism.

One of the great paradoxes of modernity is that even as scientists begin to articulate the laws that govern nature in general, the 'presentism' that goes with it marks the beginning of modern man as individual, the origin of himself as we see him portrayed in the self-inventing Robinson Crusoe. [19] He was to become the culture hero of modernity, self-created, self-narrated, existentially alone, and, as an observing subject, an outsider with respect to the laws he discovers even about himself as an object. At the same time he is the necessary channel for discovery. In the new world, tradition, old books, immanent ideas, are old superstitions. Knowledge, as Donne laments, is what I can measure and observe here and now in the things before me. With this change, the semantics of the terms originality and invention begin to change, too. 'Original' gradually comes to define what I have just found out, something never seen before; 'invention' spells the capacity for thinking new thoughts. The modern world has arrived.

18 Ibid., p. 492.
19 The term 'presentism' was introduced by Paul Hunter in *Before the Novel* (New York and London: Norton, 1992).

Jón Gunnar Þorsteinsson

Paradigm as an Origin: Or, Bearing Evidence to the Meaningful Logic of Cause and Effect

I. Origin

Origin is an important concept, both in the sciences and in the humanities. One of the most famous and influential books of science has the word origin in the title, namely Charles Darwin's *On the Origin of Species by Means of Natural Selection, or the Preservation of Favoured Races in the Struggle for Life* [1] (1859), usually referred to as *The Origin of Species*, which emphasises even more the word origin. Within the humanities, questions about origins surface easily, and they are often difficult to answer. We are normally curious about the origin of things and phenomena, for example the origin of literature, writing, of social behaviour, customs, and so forth. Such questions are interesting, perhaps because they cannot be answered properly.

Origin is a problematic concept. One could easily argue that the last four decades of critical thought have involved a deconstruction of the term origin, questioning the idea that there is a single, unified origin which gives meaning to what follows it. French post-structuralism was pivotal in this deconstruction. Roland Barthes' famous essay 'The Death of the Author' can for example be considered as one of several starting points of the deconstruction of origin. The Author, in the following quotation from Barthes' essay, is an archetypal example of origin:

> Once the Author is removed, the claim to decipher a text becomes quite futile. To give a text an Author is to impose a limit on that text, to furnish it with a final signified, to close the writing. [2]

1 Charles Darwin, *On the Origin of Species: A Facsimile of the First Edition with an Introduction by Ernst Mayr* (Cambridge: Harvard University Press, 1964).

2 Roland Barthes, 'The Death of the Author', in *Image, Music, Text: Essays Selected and Translated by Stephen Heath* (London: Fontana Press, 1977), pp. 142–48 (p. 147). Originally published as 'La mort de l'auteur', in *Mantéia*, V (1968), 12–17.

When the Author is removed, the text is freed and given to the reader, and at that point it gains a unity that cannot be found in its origin but in its destination. Texts are further described in the article 'From Work to Text', where Barthes elaborates on the binary opposition text/work. The work exemplifies literature owned by an author: 'The work is caught up in a process of filiation [...]. The author is reputed the father and the owner of his work'. [3] The text, on the other hand, lacks its traditional origin, 'it reads without the inscription of the Father' (p. 161). As a consequence, it has multiple meanings; it is not closed off by a final signified, but elicits the play of signifiers.

One of Barthes' metaphor for the text is that of a network that might be expanding 'as a result of a combinatory systematic (an image, moreover, close to current biological conceptions of the living being)' (p. 161). He describes the structure and form of the fatherless text in the following way: it has lost its origins, has no conceivable end, and is without a centre. The text 'is structured but off-centred, without closure [...] a system with neither close nor centre' (p. 159). The text as a network is an apt description, but another metaphor might also be suggested, namely the universe. One of the key facts of the universe in modern cosmology is that it is continuously expanding, and has been so since the big bang. The universe has no centre and might continue to expand forever. [4] Although modern cosmology and Roland Barthes' critical writings might seem far from each other, they provide an interesting ground for comparing humanistic and scientific discourses about the text and origins. Cosmology deals with the ultimate origin, namely the origin of the universe. Within the humanities, post-structuralism played an important role in the deconstruction of a single unified origin. These two discourses will be compared not only to highlight relationships between two seemingly different fields, but also to emphasise how a discourse about origins in science can serve as a good lesson within the humanities. Other scientific and humanistic notions of origins, such as the origin of a river, of cartoons and of Modernism, will be examined in order to enforce the main argument.

My contention is that the search for origins of phenomena and concepts can move us away from the very same concepts and phenomena. When one is looking for origins it is usually possible to go one step further back, and from that step, another one backwards. This process of tracing backwards can lead any serious analysis astray. In order to avoid the risk of undermining the cause-and-effect nexus, it is important to explore the concept of paradigm.

3 Roland Barthes, 'From Work to Text', in *Image, Music, Text*, pp. 155–164 (p. 160). Originally published as 'De l'œuvre au texte', in *Revue d'esthétique*, 3 (1971), 226–27. Further references are given after quotations in the text.

4 See for example William J. Kaufmann III and Roger A. Freedman, *Universe*, 6th edition (New York: W. H. Freeman and Company, 2002), pp. 640–61.

Origin or origins are difficult paradigms, but paradigm, on the other hand, might be a good replacement for the term origin. In Thomas Kuhn's sense, a revolution in science takes place when one paradigm replaces another. [5] Paradigms shifts are essentially representations of significant changes. The transition from the classical mechanics of Isaac Newton to Albert Einstein's relativity is a typical example. Consider a scientist working on relativity. Regarding Einstein's physics as a paradigm allows for focusing on a definite object of enquiry, while searching for the origin of the concept of relativity opens up distant fields which do not matter much to the theoretical framework. It is perhaps significant that *The Structure of Scientific Revolutions* starts almost by doubting the relevance of questions about origins:

> In recent years, however, a few historians of science have been finding it more and more difficult to fulfil the functions that the concept of development-by-accumulation assigns to them. As chroniclers of an incremental process, they discover that additional research makes it harder, not easier, to answer questions like: When was oxygen discovered? Who first conceived of energy conservation? Increasingly, a few of them suspect that these are simply the wrong sorts of questions to ask. [6]

The concept of paradigm is Kuhn's answer to what he calls the wrong questions. Accumulating material in order to address issues of origins makes the answers more difficult. The questions might in fact be the wrong ones to ask.

II. Original Meaning

According to the *Oxford English Dictionary* the meaning of origin is the following: 'The act or fact of arising or springing from something; derivation, rise; beginning of existence in reference to its source or cause'. [7] The English word *origin* derives from Middle English *origine*, probably borrowed from Middle French, itself deriving from Latin *orīgō, orīgin-*, from *orīrī*, to arise, be born. Etymology arouses the following question: is the meaning of origin in the dictionary the original meaning of the word? If one is looking for the original meaning, is it necessary to trace the word to its origin? The question can be answered in several ways. If one accepts that the English word *origin* is the same as, or equivalent to, the Middle English word *origine* and the Latin word *orīgō*,

5 Thomas S. Kuhn, *The Structure of Scientific Revolutions*, 2nd edition (Chicago: The University of Chicago Press, 1970), p. 111.
6 Kuhn, p. 2.
7 *The Oxford English Dictionary*, ed. by J. A. Simpson and E. S. C. Weiner, 20 vols, 2nd edition (Oxford: Oxford University Press, 1989), X, p. 933.

then we might not have found the 'original meaning' of the English word origin. It would then be necessary to trace it back. But if we assert that there is a difference between the English word *origin*, Middle English *origine* and Latin *orīgō*, that they are not the same words, then we probably do not need to trace the meaning of the English word origin any further back, unless perhaps to retrieve the first use of the word in English. But that might be difficult. How could the first instance of the word be dated? And why should the original meaning of the word be more important than successive meanings, or slight nuances?

Finding the 'actual' or 'true' origin of the word origin thus proves less important than answering the question: what does original meaning mean? The term is in itself highly problematic, being constructed within the binary opposition of the terms origin – conceived of as original, authentic, superior – and end. What should be remembered about etymology and origin is that by tracing words back to a presumed origin we are not uncovering more original meaning, in terms of the origin-and-end opposition, but rather uncovering differences. We can find a different meaning, and of course we might also find the same meaning.

III. Source of a River

How would one clarify the concept of origin to someone who knows very little about origins? It might for example be possible to follow a big river upstream and reach the source, which would be a good example of origin: it is the point where something begins or rises, or from which it derives. However, things are more complicated, because the real origin of the water at the origin of the river is rainfall, which evaporates from the sea, which comes originally from volcanic eruptions. It would even be possible to carry the search for the origin of a river to four and a half billion years ago, when the planet Earth was formed and the water we now see on its surface was within it. We could of course go on and arrive at the beginning of the universe. [8]

This is the reason why the origin of a river or river source has to be defined very clearly. If the concept of origin is not supported by further clarifications, it is easy to get into trouble. Searching for the origin of a river shows that we can loose sight of what we want to clarify. Instead of explaining the concept of origin, we end up giving a small lecture about the circulation of water on the Earth.

8 For a brief and accessible account of the circulation of water on Earth, see for example W. Kenneth Hamblin and Eric H. Christiansen, *Earth's Dynamic Systems*, 9th edition (New Jersey: Prentice Hall, 2001), pp. 31 – 38.

IV. Origin of Cartoon Characters

The origin of cartoon characters is another good example. The word cartoon has two main meanings. It is used for cartoons in newspapers and comic magazines, and also for a preparatory design, drawing, or painting, such as a fresco. [9] The etymology is the Italian word *cartone*. The word cartoon used for a preparatory design is thus older, or more 'original', while the first known use of the word cartoon in the other sense dates back to the year 1843, when it appeared in the British magazine *Punch*.

However, cartoon characters can be found at an earlier date. For example Dr Syntax was created in 1809 by the English caricaturist Thomas Rowlandson (1757 – 1827), one of the founders of the English caricature tradition. It could therefore be argued that the pedantic Dr Syntax is the grandfather of modern cartoon characters. [10]

Perhaps it is possible to go much further back. Consider for example the medieval illustrations of the so-called Plinian races. For the modern reader several illustrations might be considered humorous, as should be evident from John Block Friedman's description of some races:

> *Amyctyrae* ('unsociable'). This race has a lower lip – or sometimes an upper – that protrudes so far that it can serve as an umbrella against the sun [...] *Blemmyae*. In the deserts of Libya, according to Pliny, live men with their faces on their chests; they lack heads and necks [...] The remarkable ears of Ctesias' Pandae are, as it were, detached for their owners and made the attributes of a separate race called Panotii, whose ears reach to their feet and serve as blankets. The Panotii are very shy, and when they see travellers they unfurl their ears and fly away as if with wings. A fine specimen, with ears looped over his arms like sausages, appears in the Anglo-Saxon Wonders manuscript. [11]

These selected descriptions of the Plinian races fit well within the definition of the term caricature: 'Type of art in which the characteristic features of the human figure are exaggerated for amusement or criticism'. [12] It might therefore be possible to argue that for the contemporary viewer the Plinian races look like medieval forerunners of cartoons. But why stop at the Middle Ages and not go all the way back to the first known pictorial representations, i. e. cave paintings from

9 *The Oxford English Dictionary*, II, p. 927.

10 On Thomas Rowlandson and Dr Syntax see for example *The Dictionary of Art*, ed. by Jane Turner, 34 vols (New York: Grove, 1996), XXVII, pp. 277 – 79.

11 John Block Friedman, *The Monstrous Races in Medieval Art and Thought* (New York: Syracuse University Press, 2000), pp. 9 – 18.

12 *The Dictionary of Art*, V, p. 755.

prehistoric times, and argue that, by stretching the definition to its limits, they might be considered cartoons?

Definite human figures seldom appear in prehistoric parietal art. Scholars therefore subdivide depictions of people in cave art into three categories: humans, humanoids and composites. [13] The following description of the so-called sorcerer of Trois Frères fits well within the definition of cartoons:

> The best-known composite figure is the so-called sorcerer of Trois Frères, who, despite his erect stance and human legs and hands, has the back and ears of a herbivore, the antlers of a reindeer, the tail of a horse and the phallus in a feline position. [14]

Is it relevant to trace the origin of the concept of cartoon characters all the way back to the first known paintings? Is it relevant, if one wants to analyse cartoons? Indeed, such an analysis is totally blinded to history, because it does not highlight how readers of the Middle Ages regarded illustrations of the Plinian races, or what the creators of the cave paintings thought about the figures they painted. Looking for the origin of cartoons gets us away from the concept of cartoons, while considering them as a paradigm keeps us within meaningful boundaries of the concept.

V. Origin of Modernism

A common problem in historical research is deciding where one should stop tracing phenomena back, when it is irrelevant to go any further backwards. In his essay 'Thoroughly Modern Modernism: Modernism and its Postmodernism' Sam Slote problematizes the 'novelty' of Modernism by maintaining that it is possible to trace the origin of modernistic traits much further back than the period within which we are used to locate Modernism.

> Modernism is nothing new. This would be to say that certain typical tendencies and traits associated with Modernism can be readily identified in earlier works, at least with the benefit of hindsight. Self-conscious irony is hardly exclusive to Joyce, Nabokov, and Queneau, it occurs in *Don Quixote* and Lucian's *A True Story* to take but two examples. Deliberate syntactic gaps and slippages are not innovations of Pound and Rilke, for they can be seen in some of Horace's poems. [...] Twelve tone progressions could be heard long before Schönberg in Bach's *Suites for Solo Cello*. And Blake is a forerunner, of sorts, to the 'multimedia' experimentation of writers

13 Ibid., XXV, p. 481.
14 Ibid., ivi.

and artists like Wyndham Lewis and Picasso. (And, this point about the unoriginality of Modernism is also not exactly new).[15]

It is thus not fundamental to find the precise origin of Modernism, and anyhow it will not be found. The concept of paradigm is much more relevant. Indeed, if we focus on Modernism as a paradigm, instead of trying to locate its origin, the discussion becomes more productive.

Tracing the origin can become a game that transcends the logic of cause and effect. It is well known that Bach, Blake and Horace were not Modernists, although it might be possible to find modernistic traits in their work. Searching for the origin of Modernism might actually move us away from Modernism, while focusing on it as a paradigm moves us back to where it belongs.

While Kuhn's paradigm is certainly relevant in this context, Ludwig Wittgenstein's family resemblances can also point to the right direction. Wittgenstein enunciates his concept of family resemblances with regard to a discussion about the difficulties of clearly defining the concept of games:

> Consider for example the proceeding that we call 'games'. I mean board-games, card-games, ball-games, Olympic games and so on. What is common to them all?[16]

Wittgenstein's answer to the question is basically that there is no single and definable thing common to all games. Instead there is 'a complicated network of similarities overlapping and criss-crossing: sometimes overall similarities, sometimes similarities of detail'.[17] And in order to characterise these similarities Wittgenstein propounds the concept of 'family resemblances'. If one thinks about family resemblances in relation to Modernism and literature in general, it should be clear how the concept can guide us in not loosing sight of what Modernism actually means, without defining the concept too precisely, for it is always easy to argue about definitions.

The several types of games mentioned by Wittgenstein, such as board-games and card-games all belong to the category of games. Modernism is a similar kind of category, and modernistic works such as Joyce's *Ulysses*, Eliot's *The Waste Land*, Faulkner's *The Sound and the Fury* and so forth are the types. The different

15 Sam Slote, 'Thoroughly Modern Modernism: Modernism and its Postmodernisms', in *Modernism*, ed. by Ástráður Eysteinsson and Vivian Liska, 2 vols (Amsterdam and Philadelphia: John Benjamins Publishing Company, 2007), I, p. 233.

16 Ludwig Wittgenstein, *Philosophical Investigations: The German Text, with a Revised English Translation*, trans. by G. E. Anscombe, 3rd edition (Malden: Blackwell Publishing, 2001), p. 27e.

17 Ibid., ivi.

types of modernistic works do not have any single and definable thing in common, instead they all belong to this complicated network Wittgenstein talks about. But why should Horace's poems or the works of William Blake not be included in the same category, since they do also have some resemblances? It is possible to do so, but then we would be mixing up categories, namely the category of literature in general and the category of Modernism. One of the resemblances inherent in all the types belonging to Modernism is that they are rooted in history. Modernism is a historical concept.

If one is addressing literature in general, it might be perfectly acceptable to connect James Joyce's self-conscious irony to *Don Quixote*, and so forth. But if one is focusing on Modernism as a category, the discussion must be limited to Modernism as a single family. Once the modernistic family is enlarged too much with authors like Horace or texts like *Don Quixote*, meaningful relations are breached.

VI. Origin as a Problematic Concept

The concept of origin has been problematic for a long time in several academic fields. For some of them it is even a taboo. In a book called *Origins: The Darwin College Lectures*, published in 1988, seven scholars write about the origin of the universe and the solar system, human species, language, and so forth. Two of the authors declare that the concept of origin has often caused trouble in their field. Ernest Gellner, who writes about the origin of society, maintains:

> The subject which I am paid to teach in this University, Social Anthropology, has been defined for something like five or six decades by the prohibition, and that is not too strong a word, of speculation about origins. This is not just a precept built into the mores of the community which teaches the subject, but has in effect become part of its definition. [18]

Gellner provides a specific reason for his willingness to talk about the origins of society. He argues 'that prohibition of speculation about origins [...] call[s] for serious discussion and revisions' (p. 128). Gellner is therefore more interested in the prohibition of speculating about origins, than in origins themselves.

In the same book, John Lyons, who writes about the origins of language, starts his article by noting that:

18 Ernest Gellner, 'Origin of Society', in *Origins: The Darwin College Lectures*, ed. by A. C. Fabian (Cambridge: Cambridge University Press, 1988), p. 128. Further references are given after quotations in the text.

what looks like a nice set of sensible, interesting and, at least in principle, answerable questions – how, when, where and perhaps also why did language originate? – turns out to be, on examination, one that is riddled with hidden presuppositions and ambiguities. (p. 141)

Lyons also mentions that philologists and linguists who speak or write about the origin of language have a traditional and almost mandatory duty to mention a well known ban by the Linguistic Society of Paris from the year 1866. One of the Society statute was the following: 'The Society does not accept papers on either the origin of language or the invention of a universal language' (p. 142). However, Lyons does not refrain from tackling the origin of language, for he is not defining a single origin of language, but rather discussing 'multiple (and indefinite) origins' (p. 142).

Lyons and Gellner are both willing to analyse and work on the concept of origin, but it should be noted that they do not examine the concept itself. Gellner is interested in the prohibition of speculating on origins, and Lyons focuses upon multiple and indefinite origins.

VII. Origin of the Universe

Origin used to be a very problematic concept in the field of cosmology. In his book *The First Three Minutes*, first published in 1977, the American physicist and Nobel laureate Steven Weinberg writes about how difficult it was for him and other physicists in the 1950s and 60s to convince themselves that the universe had an origin. Weinberg writes:

From the start of modern science in the sixteenth and seventeenth centuries, physicists and astronomers have returned again and again to the problem of the origin of the universe. However, an aura of the disreputable always surrounded such research. I remember that during the time that I was a student and then began my own research (on other problems) in the 1950s, the study of the early universe was widely regarded as not the sort of thing to which a respectable scientist would devote his time. [19]

There were specific reasons for this disregard, because: 'Throughout most of the history of modern physics and astronomy, there simply has not existed an adequate observational and theoretical foundation on which to build a history of the early universe' (p. 4). But since then a lot has changed in the field of cos-

19 Steven Weinberg, *The First Three Minutes: A Modern View of the Origin of the Universe* (New York: Basic Books, 1977), p. 4.

mology: a theory of the early universe has become so widely accepted that astronomers often call it 'the standard model'. It is more or less the same as the 'big bang' theory. Weinberg's uncertainties about analysing the origins of the universe are very understandable. When scientists had no observational or theoretical foundation for enquiring into the origins, these matters were probably best left to scholars of philosophy and theology. The discourse about the ultimate origin in cosmology can serve as a good lesson in post-modern theory. For what I see as a problem with the concept of origin is that it is often difficult to know when or where to stop when we are searching for origins.

VIII. Origin as a Signifier with Unsteady Signifieds

In their recent book Lorraine Daston and Peter Galison maintain that objectivity in science originated in the mid-nineteenth century, although 'skeptics will perhaps be entertained but unimpressed by the curious history of the word 'objectivity'. Etymology is full of oddities, they will concede, but the novelty of the word does not imply the novelty of the thing'. [20]

According to Daston and Galison, the question thus becomes: why should we not trace the history of objectivity further back? What about Bacon and Descartes in the seventeenth century? And 'why doesn't the mathematical natural philosophy of Newton or the painstaking microscopic research of Antonie van Leeuwenhoek qualify as a chapter in the history of objectivity?' [21]

The answer to these questions is that, although possible, tracing objectivity further back would have no special meaning, because what the authors are actually discussing is objectivity as a paradigm. Reaching out of that framework bears no meaning on objectivity in the sense they are talking about. Similarly, finding modernistic traits in Horace, Bach and Blake, although possible, has probably no significant meaning for Modernism as a concept and a paradigm.

There is one good lesson we can learn from scientists within the field of cosmology: the ultimate origin might not exist. But if it does, there is no point in enquiring about it, since, as a point of origin, it does not have any meaning. As Steven Weinberg aptly argues:

However, although we do not know that it is true, it is at least logically possible that there *was* a beginning, and that time itself has no meaning before that moment. We are all used to the idea of an absolute zero of temperature. It is impossible to cool anything below -273.16 °C not because it is too hard, or because no one has

20 Lorraine Daston and Peter Galison, *Objectivity* (New York: Zone, 2007), p. 31.
21 Ibid., p. 34.

thought of a sufficiently clever refrigerator, but because temperatures lower than absolute zero just have no meaning – we cannot have less heat than no heat at all. In the same way, we may have to get used to the idea of an absolute zero of time – a moment in the past beyond which it is in principle impossible to trace any chain of cause and effect. The question is open, and may always remain open. [22]

This is essentially where Steven Weinberg agrees with Roland Barthes – where scientific cosmology meets French post-structuralism. If we relate Weinberg's remarks to Roland Barthes' post-structuralist discourse, we could say that the ultimate origin, or the final signified, are in fact unsteady. Those who think that the Author will help them find the final signified – the ultimate meaning – of the text might in the end only encounter a loss of meaning. They are asking the wrong questions, as Kuhn says.

Origins, in the sense that there is a precise beginning more meaningful than what follows, is a false one. Focusing on origin as a concept might lead us to undermine the chain of cause and effect. Although we can often trace the origin of phenomena much further back, the process can become pointless. At risk is that we get lost in arguing about definitions which are perhaps not relevant in themselves – instead of doing work on the actual phenomena. Paradigm, on the other hand, is a word where the chain of cause and effect has a meaning.

22 Weinberg, pp. 148–49.

Bibliography Part III

Alford, H., ed., *The Works of John Donne, D. D., Dean of Saint Paul's, 1621 – 1681. With A Memoir of His Life*, 6 vols (London: John W. Parker, 1839).

Assmann, J., *Das kulturelle Gedächtnis. Schrift, Erinnerung und politische Identität in frühen Hochkulturen* (München: Beck, 1992).

Barthes, R., 'The Death of the Author', in *Image, Music, Text: Essays Selected and Translated by Stephen Heath* (London: Fontana Press, 1977), first published 1968, pp. 142 – 48.

Barthes, R., 'From Work to Text', in *Image, Music, Text: Essays Selected and Translated by Stephen Heath* (London: Fontana Press, 1977), first published 1971, pp. 155 – 64.

Birrel, A., *Chinese Mythology. An Introduction* (Baltimore: John Hopkins University Press, 1993).

Bock, J., *Skipholt* (Köln: Verlag der Buchhandlung Walther König, 2008).

Boisacq, E., *Dictionnaire étymologique de la langue grecque* (Heidelberg: Winter, 1923).

Børch, M., 'Chaucer and the Cosmic Text: Rejecting Analogy', in *Text and Voice: The Rhetoric of Authority in the Middle Ages*, ed. by M. Børch (Odense: University Press of Southern Denmark, 2004), pp. 97 – 120.

Børch, M., 'Chaucer's Second Nun's Tale: Record of a Dying World', in *A Chaucer Yearbook 5*, ed. by J. Jost and M. Salda (Brewer: Cambridge, 1999), pp. 19 – 40.

Børch, M., 'Love's Ontology and the Problem of Cliché', in *Sponsored by Demons: The Art of Jeanette Winterson*, ed. by H. Bengtson, M. Børch, and C. Maagaard (Odense: Scholars' Press, 1999), pp. 41 – 54.

Brewer, D., 'Chaucer's Poetic Style', in *The Cambridge Chaucer Companion*, ed. by P. Boitani and J. Mann (Cambridge: Cambridge University Press, 1986), pp. 227 – 42.

Brewer, L. C., ed., *The Riverside Chaucer* (London: Oxford University Press, 1988).

Brisson, L., *Orphée et l'Orphisme dans l'antiquité gréco-romaine* (Aldershot: Ashgate, 1995).

Burkert, W., *Da Omero ai Magi. La tradizione orientale nella cultura greca* (Venezia: Marsilio, 1999).

Bury, R. G., ed., *Plato. Timaeus, Critias, Cleitophon, Menexenus, Epistles* (Cambridge – London: Harvard University Press, 1929).

Campbell, A., ed., *The Battle of Brunanburh* (London: William Heinemann Ltd, 1938).

Carruthers, M., *The Book of Memory* (Cambridge: Cambridge University Press, 1992).

Cavalli, E., *Il divano del Nord. Viaggio in Scandinavia* (Milano: Feltrinelli, 2005).

Cavazza, F., *Lezioni di indoeuropeistica con particolare riguardo alle lingue classiche (sanscrito, greco, latino, gotico)* (Pisa: Ets, 2001).
Chantraine, P., *Dictionnaire étymologique de la langue grecque. Histoire des mots* (Paris: Klincksieck, 1980).
Cleasby, R., G. Vigfusson, and W. A. Craigie, *An Icelandic – English Dictionary*, 2nd edn (Oxford: Clarendon Press, 1962).
Codrignani, G., 'L'*aition* nella poesia greca prima di Callimaco', *Convivium*, n.s., 26 (1958), 227–45.
Colish, M., *The Mirror of Language: A Study in the Medieval Theory of Knowledge* (New Haven: Yale University Press, 1968).
Cooper, H., *The Structure of the Canterbury Tales* (London: Duckworth, 1983).
Coumert, M., *Origines des Peuples. Les Récits du Haut Moyen Âge Occidental* (550–850) (Paris: Institut d'Études Augustiniennes, 2007).
Dalley, S., *Myths of Mesopotamia* (Oxford: Oxford World's Classics, 1989).
Darwin, C., *On the Origin of Species: A Facsimile of the First Edition with an Introduction by Ernst Mayr* (Cambridge: Harvard University Press, 1964).
Daston, L., and P. Galison, *Objectivity* (New York: Zone, 2007).
De Michelis, M., *Il sublime è ora / The Sublime is Now* (Milano: Skira, 2008).
Dumville, D. N., 'The Anglian Collection of Royal Genealogies and Regnal Lists', *Anglo-Saxon England*, 5 (1976), 23–50.
Dronke, U., ed., *The Poetic Edda*, 2 vols (Oxford: Clarendon Press, 1997).
Egner, T., *Karius og Baktus* (Oslo: Gyldendal, 2001; 1st edn 1949). English version: <http://www.youtube.com/watch?v=VOJo73IGpfE> [accessed 3 September 2009].
Elton, O., ed., *Saxo Grammaticus. The First Nine Books of the Danish History* (London: David Nutt, 1894).
Erdman, D. V., ed., *The Complete Poetry and Prose of William Blake*, with a New Foreword and Commentary by H. Bloom (Berkley and Los Angeles: University of California Press, 2008 [first publ. 1965]).
Fabian, A. C., ed., *Origins: The Darwin College Lectures* (Cambridge: Cambridge University Press, 1988).
Faulkes, A., ed., *Snorri Sturluson. Edda* (London: Everyman, 1997).
Frazer, J., ed., *Apollodorus. The Library*, 2 vols (Cambridge, Mass.: Harvard University Press 1976–1979).
Friedman, J. B., *The Monstrous Races in Medieval Art and Thought* (New York: Syracuse University Press, 2000).
Frisk, H., *Griechisches Etymologisches Wörterbuch*, 3 vols, 2nd edn (Heidelberg: Carl Winter Universitätsverlag, 1973).
Garmonsway, G. N., ed., *The Anglo-Saxon Chronicle* (London: J. M. Dent & Sons Ltd., 1972).
Gellrich, J., *The Idea of the Book in the Middle Ages* (Ithaca: Cornell, 1985).
Gransden, A., *Historical Writing in England c. 550 to c. 1307* (London: Routledge & Kegan Paul, 1974).
Groningen, B. A. van, *In the Grip of the Past. Essay on an Aspect of Greek Thought* (Leiden: E. J. Brill, 1953).
Halbwach, M., *La mémoire collective* (Paris: Presses universitaires de France, 1968; 2. éd. revue et augmentée).

Hamblin, W. K., and E. H. Christiansen, *Earth's Dynamic Systems*, 9th edition (New Jersey: Prentice Hall, 2001).

Hauck, K., 'Lebensnormen und Kultmythen in germanischen Stammes- und Herrschergenealogien', *Saeculum*, 6 (1955), pp. 186 – 223.

Henderson, J., ed., Aristophanes. *Birds, Lysistrata, Woman at the Thesmophoria*, (Cambridge – London: Harvard University Press, 2000).

Herrenschmidt, C., 'Entre Perses et Grecs. I. Démocrite et le mazdéisme: Religion, philosophie, science', *Transeuphratène*, 11 (1996), 115 – 43.

Higham, N. J., *An English Empire. Bede, the Britons and the Early Anglo-Saxon Kings* (Manchester – New York: Manchester University Press, 1995).

Hoffner, H. A., and G. M. Bekman, *Hittite Myths* (Atlanta: Scholars Press, 1998).

Hollander, L. M., ed., *The Poetic Edda* (Austin: University of Texas Press, 1962).

Hunter, P., *Before the Novel* (New York and London: Norton, 1992).

Huyghe, P., *One Million Kingdoms* (2001), in C. Christov-Bakargiev, *Pierre Huyghe* (Skira: Genève-Milano, 2004).

Johnson, E. G., ed., *Sir Joshua Reynolds's Discourses*, with Illustrations (Chicago: A. C. McClurg and Company, 1891).

Jones, W. H. S., ed., *Pausanias Periegeta. The Description of Greece*, 5 vols (Cambridge, Mass.: Harvard University Press, 1965 – 1978).

Kaufmann, W. J. III, and R. A. Freedman, *Universe*, 6th edition (New York: W. H. Freeman and Company, 2002).

Kern, O., ed., *Orphicorum Fragmenta* (Berlin: Weidmann, 1922).

Keynes, S., and M. Lapidge, eds, *Alfred the Great: Asser's Life of Alfred and Other Contemporary Sources* (London: Penguin, 1983).

Klaeber, F., ed., *Beowulf and the Fight at Finnsburg* (Lexington, Massachusetts: D. C. Heath & Company, 1950).

Kleingünter, A., *Protos Heuretes: Untersuchungen zur Geschichte einer Fragestellung* (Leipzig: Akademie Verlag, 1933).

Kuhn, T. S., *The Structure of Scientific Revolutions*, 2nd edition (Chicago: The University of Chicago Press, 1970).

Laks, A., and G. W. Most, *Studies on the Derveni Papyrus* (Oxford: Clarendon Press, 1997).

Lamb, W. R. M., ed., *Plato. Laches, Protagoras, Meno, Eutydemus* (Cambridge – London: Harvard University Press, 1924).

Lehmann, W. P., *A Gothic Etymological Dictionary* (Leiden, E. J. Brill, 1986).

Lichtheim, M., *Ancient Egyptian Literature*, 3 vols (Berkeley: The University of California Press, 1973 – 1980).

Lincoln, B., 'The Center of the World and the Origins of Life', *History of Religions*, 40 (2001), 311 – 26.

Maggia, F., and S. Ferrari, eds, *Northern Lights. Reflecting with Images* (Cinisello Balsamo: Silvana Editoriale, 2007).

Mann, J., 'Malory: Knightly Combat in Le Morte d'Arthur', in *The New Pelican Guide to English Literature*, 8 vols, ed. by B. Ford (Harmondsworth: Penguin, 1982), I. Medieval Literature. Chaucer and the Alliterative Tradition, pp. 332 – 35.

Marcovich, M., ed., Hyppolitus of Rome, *Refutatio omnium heresium* (Berlin – New York: Walter de Gruyter, 1986).

Mayr-Harting, H., *The Coming of Christianity to Anglo-Saxon England* (London: Batsford, 1972).

Meli, M., *Völuspá. Un'apocalisse norrena* (Roma: Carocci, 2008).

Michelet, F. L., *Creation, Migration, and Conquest. Imaginary Geography and Sense of Space in Old English Literature* (Oxford: Oxford University Press, 2006).

Moisl, H., 'Anglo-Saxon Royal Genealogies and Germanic Oral Tradition', *Journal of Medieval History*, 7 (1981), 215 – 48.

Monier-Williams, M., *A Sanskrit – English Dictionary Etymologically and Philologically arranged with special reference to cognate Indo-European Languages* (Dehli: Motilal Banarsidass Publishers, 1995).

Most, G. W., ed., *Hesiod. Theogony, Works and Days, Testimonia* (Cambridge – London: Harvard University Press, 2006).

Murray, A. T., ed., *Homer. The Iliad*, 2 vols (London, W. Heinemann – Cambridge, Mass., Harvard University Press, 1934 – 37).

Neckel, G., and H. Kuhn, eds, *Edda. Die Lieder des Codex Regius nebst verwandten Denkmälern*, 5th edn (Heidelberg: Carl Winter Universitätsverlag, 1983).

Nelson, J. L., 'Reconstructing a Royal Family: Reflections on Alfred, from Asser, chapter 2', in *People and Places in Northern Europe, 500 – 1600*, ed. by I. Wood and N. Lund (Woodbridge: Boydell, 1991), 47 – 66.

Origene [Origenes], *Commento al Vangelo di Giovanni*, ed. by E. Orsini (Torino: UTET, 1968).

Osborne, H., *Aesthetics and Art History* (New York: Dutton, 1970).

Osgood, Ch. W., ed., *Boccaccio on Poetry: Being the Preface and the Fourteenth and Fifteenth Books of Boccaccio's 'Genealogia Deorum Gentilium'* (Indianapolis and New York: Liberal Arts Press, 1956).

Pagani, I., ed., *Adamo di Brema. Storia degli arcivescovi della Chiesa di Amburgo* (Torino: UTET, 1996).

Pazzini, D., *In principio era il Logos. Origene e il prologo del Vangelo di Giovanni* (Brescia: Paideia, 1983).

Penglase, C., *Greek Myths and Mesopotamia: Parallels and Influence in the Homeric Hymns and Hesiod* (London: Routledge, 1994).

Plassmann, A., *Origo Gentis. Identitäts- und Legitimitätsstiftung in früh- und hochmittelalterlichen Herkunftserzählungen* (Berlin: Akademie Verlag, 2006).

Plummer, C., and J. Earle, eds, *Two of the Saxon Chronicles Parallel* (Oxford: Oxford University Press, 1929).

Pokorny, J., *Indogermanisches Etymologisches Wörterbuch*, 4th edn, 2 vols (Tübingen and Basel: A. Francke Verlag, 2002).

Pritchard, J. B., *Ancient Near Eastern Text Relating to the Old Testament* (Princeton: Princeton University Press, 1950).

Richardson, N., ed., *Homeric Hymns*, trans. by J. Cashford (London – New York: Penguin, 2003).

Ricci, P. G., ed., *Dante Alighieri. De Vulgari Eloquentia* (Firenze: Le Monnier, 1968).

Roberts, A., J. Donaldson, and A. C. Coxe, eds, *Ante-Nicene Fathers*, 10 vols (Buffalo, NY: Christian Literature Publishing Co., 1886).

Robertson, D. W., *A Preface to Chaucer* (Princeton: Princeton University Press, 1962).

Schützeichel, R., *Althochdeutsches Wörterbuch*, 4th edn (Tübingen, Max Niemeyer, 1989).

Schwyzer, E., *Griechische Grammatik im Anschluss an Karl Brugmanns griechische Grammatik. Allgemeiner Teil und Lautlehre* (München: C. H. Beck'sche Verlagsbuchhandlung, 1934).
Scragg, D. C., 'Secular Prose', in *A Companion to Anglo-Saxon Literature*, ed. by P. Pulsiano and E. Treharne (Oxford: Blackwell, 2001), pp. 268 – 80.
Sherley-Price, L., ed., *Bede. A History of the English Church and People*, rev. by R. E. Latham (Harmondsworth: Penguin Books, 1968).
Simpson, J. A., and E. S. C. Weiner, eds, *The Oxford English Dictionary*, 20 vols, 2nd edition (Oxford: Oxford University Press, 1989).
Sims-Williams, P., 'The Settlement of England in Bede and the *Chronicle*', *Anglo-Saxon England*, 12 (1983), pp. 1 – 41.
Sisam, K., 'Anglo-Saxon Royal Genealogies', *Proceedings of the British Academy*, 39 (1953), 297 – 332.
Slote, S., 'Thoroughly Modern Modernism: Modernism and its Postmodernisms', in *Modernism*, ed. by Á. Eysteinsson and V. Liska, vol. 1 (Amsterdam and Philadelphia: John Benjamins Publishing Company, 2007), pp. 233 – 49.
Smith, A., *The Ethnic Revival* (Cambridge: CUP, 1982).
Smyth, A. P., *King Alfred the Great* (Oxford: Oxford University Press, 1995).
Smith, J. A., and W. D. Ross, eds, *The Works of Aristotle*, 11 vols (Oxford: Clarendon Press, 1908 - 1952).
Stevenson, W. H., ed., *Asser. Life of King Alfred* (Oxford: Clarendon Press, 1904; repr. 1959).
Swanton, M. J., ed., *The Anglo-Saxon Chronicle* (New York: Routledge, 1998).
Thesaurus Linguae Latinae (Leipzig: Teubner, 1990).
Tesky, G., ed., *John Milton. Paradise Lost* (Norton: London & New York, 2004).
Turner, J., ed., *The Dictionary of Art, 34 vols* (New York: Grove, 1996).
Vegetti, M., 'Le scienze della natura e dell'uomo nel V secolo', in L. Geymonat, *Storia del pensiero filosofico e scientifico*, 9 vols (Milano: Garzanti, 1975), I, pp. 110 – 50.
Verne, J., *Journey into the Interior of the Earth*, trans. by F. A. Malleson (London: Ward Lock & Co., 1877), http://jv.gilead.org.il/wolcott/CE_allc/ [accessed 3 September 2009].
Wawn, A., *The Vikings and the Victorians. Inventing the Old North in Nineteenth-Century Britain* (Cambridge: Brewer, 2000).
Weinberg, S., *The First Three Minutes: A Modern View of the Origin of the Universe* (New York: Basic Books, 1977).
Wenskus, R., *Stammesbildung und Verfassung. Das Werden der frühmittelalterlichen Gentes* (Köln - Wien: Bohlau, 1961).
West, M. L., '*Ab Ovo.* Orpheus, Sanchuniaton, and the Origins of the Ionian World Model', *Classical Quarterly*, 44 (1994), 289 – 307.
West, M. L., *Early Greek Philosophy and the Orient* (Oxford: Clarendon Press, 1971).
West, M. L., *Indo-European Poetry and Myth* (Oxford: Oxford University Press, 2007).
West, M. L., ed., *Hesiod. Theogony* (Oxford: Clarendon Press, 1966).
Winterson, J., *Written on the Body* (London: Vintage, 1992).
Wittgenstein, L., *Philosophical Investigations: The German Text, with a Revised English Translation*, trans. by G. E. Anscombe, 3rd edition (Malden: Blackwell Publishing, 2001).

Wormald, P., ed., *Ideal and Reality in Frankish and Anglo-Saxon Society: Studies Presented to J. M. Wallace-Hadrill* (Oxford: Blackwell, 1984).

Zironi, A., 'Il sogno premonitore dell'albero nella tradizione germanica medievale', in *Filologia romanza, filologia germanica: intersezioni e diffrazioni*, ed. by A. M. Babbi and A. Cipolla (Verona: Fiorini, 1997), pp. 343–65.

Zumthor, P., *La Mesure du Monde: Représentation de l'Espace au Moyen Age* (Paris: Editions du Seuil, 1993).

Bio-bibliographical Profiles

Alessandro Achilli received a BS at the University of Perugia and a PhD at the University of Pavia. He is assistant professor at the University of Perugia, Italy, where he teaches different courses in the field of Genetics. He is also a research assistant at the University of Pavia, Italy. He is the author and co-author of numerous publications on mitochondrial DNA variation in humans and animals with the objective to better defining population migrations, origins, and patterns of dispersal.

Sergio Agnoli is PhD in Developmental Psychology and works at the EIL (Early Infancy Lab), University of Ferrara. He is specialized in the study and analysis of infant emotional psychophysiological indexes, and emotional facial behaviour. He is an expert in neonatal and infant behaviour coding methodologies, such as Baby FACS.

Laura Bafile is Associate Professor of General Linguistics at the University of Ferrara. Her research focuses on theoretical linguistics and phonology. In particular, she investigates the nature and the internal structure of sounds, the syllable and the cognitive organisation of phonological information. She has published on issues in phonological theory, especially related to Italian varieties. She is the author of *Introduzione allo studio della grammatica* (Tecom: Ferrara, 2004), and *I suoni del linguaggio* (Bologna: Il Mulino, 2008, with Marina Nespor).

Marianne Børch is Professor of English at the University of Southern Denmark and a member of the ACUME2 Steering Committee. Her central area of research is early modern literature in English, with special reference to Chaucer, and the way literature and the other arts reflect and promote epistemological change. She is the editor of *Narratives of Remembrance* (Odense: University Press of Southern Denmark, 2001) and *Text and Voice: The Rhetoric of Authority in The Middle Ages* (Odense: University Press of Southern Denmark, 2004).

Marco Dondi is Associate Professor of Developmental Psychology and directs the EIL (Early Infancy Lab) at the University of Ferrara. He is an expert in early human emotional and cognitive development and in behavioural coding systems for infant observation. He is the author of national and international scientific publications, among which 'A New Perspective on Neonatal Smiling: Differences Between the Judgments of Expert Coders and Naïve Observers', *Infancy*, 12, 3 (2007), 235 – 55 (with D. S. Messinger, Marta Colle, Alessia Tabasso, Francesca Simion, Beatrice Dalla Barba, and Alan Fogel); 'Can Newborns Discriminate Between Their Own Cry and the Cry of Another Newborn Infant?', *Developmental Psychology*, 35, 2 (1999), 418 – 26 (with Francesca Simion and Giovanna Caltran).

Marcella Farioli has a PhD in Classical History and Philology and is a teacher of Classics in a secondary school. Her research focuses on archaic Greek comedy, Greek religion and the history of utopia in ancient times. She has published *Le religioni misteriche* (Milano: Xenia, 1998), *Mundus Alter. Utopie e distopie nella commedia greca antica* (Milano: Vita e Pensiero, 2001), and a chapter on 'Antiquité classique' in *Histoire transnationale de l'utopie littéraire et de l'utopisme*, ed. by Vita Fortunati, Paola Spinozzi and Raymond Trousson (Paris: Champion, 2008).

Vita Fortunati is Professor of English Literature at the University of Bologna. Her research focuses on modernism, Utopian Studies, Gender and Women's studies, and aging. She is the editor of *Histoire transnationale de l'utopie littéraire et de l'utopisme* (Paris: Champion, 2008, with Raymond Trousson and Paola Spinozzi); *Biocomplexity: At the Cutting Edge of Physics, System Biology and Humanities* (Bologna: BUP, 2008, with Gastone Castellani, Claudio Franceschi and Elena Lamberti); *Memories and Representations of War. The Case of World War I and World War II* (Amsterdam and New York: Rodopi, 2009, with Elena Lamberti). She has coordinated three European Thematic Networks: on *Comparative Studies* (1998 – 2002); on *Cultural Memory in European Countries* (2003 – 2006); on *Interfacing Sciences, Literature and the Humanities* (2006 – 2009).

Laura Franchin, PhD, is Research Fellow at the Department of Scienze Umane, University of Ferrara. Her main research interests lie within the area of developmental psychology, in particular neuropsychology of emotions, and emotional facial behaviour. She is a certified FACS coder and is specialized in neonatal and infant behaviour coding methodologies.

Marusca Francini has a PhD in Germanic Philology at the University of Florence and is Senior Lecturer in Germanic Philology at the University of Pavia, Italy. Her main scientific interests are the Gothic Bible, Langobardic culture and history, and Old Norse literature. She has translated and edited the *Saga di Björn, campione dei valligiani di Hítardalr* (Padova: Unipress, 2004) and is currently preparing a critical edition with commentary of the Gothic Gospel of John preserved in the *Codex Argenteus Upsaliensis.*

Andrea Grignolio, Ph.D. in history of science, focuses his research on the history of biomedical disciplines in twentieth-century western society, with special regard to immunology. In 2004 he studied at the Centre Cavaillès de l'École Normale Supérieure in Paris, in 2006–2007 was a Post-doctoral Fellowship at the Center for Philosophy and History of Science, Boston University, and in 2009 was a visiting scholar at the Office for History of Science and Technology, UC Berkeley. Recipient of the ETNP Acume2 grant, he is currently working both at the Centre 'L. Galvani', University of Bologna, and at the University of Rome 'La Sapienza'.

Keith Hoskin is Professor of Strategy and Accounting at Warwick Business School. He writes and researches across the fields of accounting, strategy, management and professionalization, but from a perspective informed by an initial background in historical, literary, philosophical and translation studies. He seeks to draw on this mix of interests to promote a greater reciprocal understanding of the continual interweaving of philosophical and historical understandings with understandings circulating within conventional human science and managerial knowledge arenas.

Ugo A. Perego, MS, is director of operations and study research coordinator for the non-profit Sorenson Molecular Genealogy Foundation (SMGF) in Salt Lake City, Utah. With SMGF he has nearly ten years of international experience in gathering biological samples, lecturing, and publishing in the field of DNA and ancestry. He is also a PhD student in human genetics at the University of Pavia, Italy, where he is studying the origins and migrations of Native American populations through the complete sequence of mitochondrial genomes. Previously, Ugo Perego earned a BS and MS at Brigham Young University in Provo, Utah.

Suresh I. S. Rattan, Ph.D., D.Sc., is Professor of Biogerontology at the Department of Molecular Biology, Aarhus University, Denmark. He has published more than 180 scientific articles and 12 books, including those for school children, general public and research scientists. His original research and areas of expertise include human cellular ageing, gerontogenes, and ageing intervention, prevention and therapies, including modulation through growth factors and mild stress induced hormesis. He is the Editor-in-Chief of *Biogerontology*, a peer reviewed international journal on the biology of ageing.

Paola Spinozzi is Senior Lecturer in English Literature at the University of Ferrara. Her interest in Interart Studies focuses on theories and methodologies, ekphrasis, illustration and calligraphy; her research in Utopian Studies covers art, aesthetics, imperialism, racism and Darwinism. She is the author of *Sopra il reale. Osmosi interartistiche nel Preraffaellitismo e nel Simbolismo inglese* (Firenze: Alinea, 2005), the editor of *Histoire transnationale de l'utopie littéraire et de l'utopisme* (Paris: Champion, 2008, with Vita Fortunati and Raymond Trousson) and *The Germ. Thoughts towards Nature in Poetry, Literature and Art* (Trento: Editrice Università degli Studi di Trento, 2008, with Elisa Bizzotto), and the co-founder, with David Skilton and Julia Thomas, of *JOIS. Journal of Illustration Studies.*

Jón Gunnar Þorsteinsson earned a B.A. in comparative literature at the University of Iceland and an M.A. in English Renaissance literature at the University of Sussex. He is assistant editor of the Icelandic Web of Science, http://www.why.is/, which aims at raising social awareness of science among the general public. He has published on early modern cultural history, early modern travel and pilgrimage literature, twentieth-century art movements, and on the audience of horror movies. He is also the editor, with Þorsteinn Vilhjálmsson, of *Af hverju er himininn blár? Spurningar og svör af Vísindavefnum* (Reykjavík: Heimskringla, 2003).

Augusto Vitale, Ph.D. in Behavioural Ecology from Aberdeen University, is a researcher in animal behaviour at the Italian National Institute of Health in Rome, Department of Cell Biology and Neuroscience. Since 1989 he has collaborated with Elisabetta Visalberghi (CNR, Rome) and Dorothy M. Fragaszy (University of Georgia), studying primates' behaviour. He has been President of the Italian Association of Primatology and is currently General Secretary for the European Federation of Primatology. His main interests are the mechanisms and evolutionary consequences of social learning, and the different aspects involved in animal experimentation, ranging from scientific to ethical factors.

Alessandro Zironi is Associate Professor of Germanic Philology at the University of Bologna. His primary areas of research include Gothic and Middle-high German languages and literatures, the historical, cultural and sociological contexts relevant to the production and circulation of Germanic texts, and codicological aspects. He has published on the reception and rewriting of medieval Germanic culture, old Norse culture and the idea of North in modern and contemporary age. Recent publications include: 'Laxdale as William Morris's Interior Topography', in *The Cultural Reconstruction of Places*, ed. by Ástráður Eysteinsson (Reykjavik: University of Iceland Press, 2006); *L'eredità dei Goti: testi barbarici in età carolingia* (Spoleto: Fondazione centro italiano di studi sull'alto medioevo, 2009).

Index

Abel, M. H. 135, 181
Achilli, A. 17, 25, 41, 55 – 57, 105, 275
Acusilaus 201
Adam 254
Adam of Bremen 218 – 219
Ades 198
Ælle 233 – 234
Aerts, J. 144, 185
Æsc 233
Agnoli, S. 17, 116, 131, 140, 182, 275
Agrippa 212
Aiken, N. E. 64, 69, 70, 75, 77, 105
Alain 66
Alalcomeneus 206
Alford, H. 254, 269
Alfred the Great (king) 229, 232 – 233, 236 – 240
Ali, R. E. 38, 105
Alighieri, D. 245
Alison 251 – 252
Ambadar, Z. 145, 186
Ambrosius Aurelianus 231
Aminocles of Corynthus 208
Ammermann, A. J. 53, 105
Amorim, A. 41, 105
Anaximander 202 – 204
Anaximenes 202 – 204
Anderson, S. 42, 105
Andonotopo, W. 145, 187
Andrews, P. 43, 111
Andrews, R. M. 42, 105
Anlaf 237
Annlee 223 – 224
Anscombe, G. E. 263
Anu 199
Apollo 209
Apollodorus 207
Apsu 198
Aristeus 208
Aristophanes 200
Aristotle 13, 61, 202 – 203
Armstrong, N. 224
Arneios 209
Aspasia 206
Asser 238
Assmann, J. 227, 269
Atys 58
Augustine (saint) 230 – 231
Augustine of Hyppon 246, 248
Avital, T. 64 – 65, 75, 105
Azumendi, G. 145, 187

Babbi, A. M. 214, 274
Bach, J. S. 262, 263, 266
Bachtin, M. 72
Bacon, F. 266
Bafile, L. 17, 117, 147, 275
Baker, M. C. 153, 181
Balda, R. P. 122, 181
Ball, S. 171, 181
Bandelt, H.-J. 41, 105
Barlow, N. 87, 105
Barner, D. 157, 184
Barthes, R. 257 – 258, 267, 269
Bartocci, M. 145, 181
Bar-Yosef, O. 52, 106
Battaglia, B. 85, 106
Bayer, R. 67

Bede 192, 229, 231 – 232, 235 – 236, 238 – 241
Bedia, J. 9, 77, 80
Beer, G. 83, 106
Beghi, A. 140, 185
Beintema, D. 142, 186
Bekman, G. M. 271
Bengston, H. 244, 269
Berenice 209
Bernes, H. E. 16
Bessenyei, B. 34, 106
Bichat, M. F. X. 173
Bickerton, D. 149, 161, 181
Bieda 234
Bignami, M. L. 85, 106
Birrel, A. 269
Bizzotto, E. 278
Blake, W. 249, 262, 264, 266
Blanc, P. 9, 75
Bloom, H. 249, 270
Bloom, P. 148 – 150, 186, 249, 270
Boccaccio, F. 245
Bock, J. 223, 269
Boisacq, E. 213, 269
Boitani, P. 250, 269
Bolzani, L. 136
Bonaparte, N. 174
Børch, M. 18, 193, 243 – 244, 253, 269, 275
Botha, R. 161, 181
Bottiroli, G. 72, 73
Bowler, P. J. 98, 106
Brancusi, C. 9, 76, 78
Brazelton, T. B. 142, 181
Breuer, H.-P. 93, 106
Brewer, D. 250, 269
Brewer, L. C. 251, 269
Brisson, L. 269
Brock, W. 16
Brosnan, S. F. 126 – 127, 181, 186
Brotherton, P. N. 123, 182
Buffon, J.-L. 28, 82, 88, 98
Bulwer Lytton, E. G. 96, 106
Burkart, J. M. 126, 181
Burkert, W. 197 – 200, 204, 269
Bury, R. G. 207, 269
Butler, S. 9, 27, 81 – 90, 92 – 103, 106
Callimachus 209
Caltran, G. 276
Campbell, A. 237, 269
Cann, R. L. 43, 106
Carnes, B. A. 32, 106
Carruthers, M. 247, 269
Cashford, J. 206, 272
Castellani, G. 276
Cavalli, E. 223, 269
Cavalli-Sforza, L. L. 47, 53 – 54, 105, 106, 109
Cavazza, F. 211, 270
Cealwin 234
Cealwine 239
Céard, J. 99, 109
Cecilia (saint) 251 – 252
Cecrops 208
Ceolwulf 235
Cerdic 234 – 235, 238 – 239
Chalmers, F. G. 66, 106
Chanda, M. L. 128, 185
Chandler, A. D. 118, 166 – 167, 177 – 178, 181
Chantraine, P. 214, 270
Chaucer, G. 193, 251
Cheney, D. L. 124, 181
Chirac, J. 74
Chomsky, N. 148 – 151, 153 – 156, 159 – 161, 169, 181, 183 – 185
Christ 246
Christensen, K. 34, 108
Christiansen, E. H. 260, 271
Christiansen, M. H. 150, 182
Christov-Bakargiev, C. 224, 271
Chronos 201
Cipolla, A. 214, 274
Cissa 234
Clarck, J. H. 9, 78
Cleasby, R. 214, 270
Clutton-Brock, T. H. 123, 182
Codrignani, G. 209, 270
Cohen, P. C. 172, 182
Cohn, J. F. 145, 182, 186 – 187
Coinmail 234
Cole, G. D. H. 88, 106
Cole, J. K. 135, 182
Colish, M. 248, 270
Colle, M. 140, 182, 276

Collingwood, R. G. 66
Collins, W. A. 139, 182
Condidan 234
Condillac, E. B. 148 - 149
Constantine 237
Cooper, H. 246, 270
Cooper, J. M. 106
Coote, J. 106
Copernicus, N. 246, 253
Corchia, C. 140, 182
Cordaux, R. 48, 106
Costabile, A. 140, 182
Coumert, M. 229, 270
Coxe, A. C. 206, 272
Crager, S. E. 128, 185
Craigie, W. A. 214, 270
Cremer, T. 32, 108
Crick, A. 16
Cronos 196
Cronus 201
Crusoe, R. 255
Ctesias 261
Ctonia 201
Culicover, P. W. 160, 182
Cutha 234
Cuthbertson, R. A. 133, 182
Cuthwine 234
Cuþwine 239
Cutwulf 234
Cymen 233
Cynric 234, 238 - 239

Dalla Barba, B. 140, 182, 276
Dalley, S. 270
Damasio, A. R. 131, 182
Damocles 172
Darwin, C. 28, 82 - 83, 86 - 89, 93 - 97, 100, 103, 107, 120, 122 - 123, 129, 132 - 133, 157, 170, 182, 257, 270
Darwin, E. 28, 82, 88, 98, 100
Darwin, F. 100
Daston, L. 266, 270
Davidson, R. J. 136, 137, 183 - 184
Davis, W. 63 - 64, 107
Dawkins, R. 86, 93, 107
De Michelis, M. 223, 270
Degueldre, C. 144, 185
Deleuze, G. 165, 182
Delfiore, G. 144, 185
Descartes, R. 266
Dessault, P. 173
Dessoir, M. 66
Diaulus 206
Dickson, K. L. 137, 139, 185
Dionysos Zagreus 208
Dissanayake, E. 64, 67 - 70, 75, 107
Dolson, B. 66, 107
Donaldson, J. 206, 272
Dondi, M. 17, 116, 131, 136, 140, 182, 185, 276
Donne, J. 254 - 255
Dougherty, L. M. 133, 185
Dracon 208
Dronke, U. 218 - 219, 270
Drury, E. 254
Drysdale, G. 96, 107
Duchenne de Boulogne, G. B. 116, 132 - 133, 136 - 137, 139 - 143, 182
Dumas, G. 135, 182
Dumville, D. N. 238, 270
Duncan, I. 96, 107
Dupanloup, I. 54, 107
Dutton, D. 59, 107

Earle, J. 232 - 233, 239, 272
Eco, U. 251
Edgell, B. 98, 107
Efferson, C. 126, 181
Egner, T. 244, 270
Eibl-Eibesfeldt, I. 121, 182
Einstein, A. 11, 259
Ekman, P. 133, 135 - 136, 139, 143, 183 - 184, 186
Eleusis 206
Eliot, T. S. 263
Elliot, G. sir 176
Elton, O. 220, 270
Emde, R. N. 137 - 138, 183
Enrich, J. 127, 186
Epimenides 201
Epimetheus 207
Erdman, D. V. 249, 270

Erebos 14
Erebus 196
Erictonius 208
Eros 196, 200
Espinoza, J. 145, 184
Ethelstan 237
Euripides 209
Eyck, J. 250
Eyer, J. 30
Eysteinsson, A. 263, 279
Ezzamel, M. 165, 183

Fabian, A. C. 12, 264, 270
Farhat, M. 73 - 74, 107
Farinmail 234
Farioli, M. 18, 191, 195, 276
Faulkes, A. 218, 270
Fehr, E. 126, 181
Fernández-Dols, J. M. 132, 186
Ferrari, C. K. B. 38, 107
Ferrari, S. 222, 271
Finch, C. E. 29, 107
Fischer, A. 131, 185
Fitch, T. W. 154, 156, 159 - 160, 183 - 185
Flacke, M. 90, 107
Fogel, A. 135 - 137, 139 - 140, 182 - 183, 185, 276
Forster, P. 48, 107
Fortunati, V. 17, 27, 81, 85, 106, 276, 278
Foss, B. M. 138, 183
Foucault, M. 118, 163, 166 - 167, 169 - 172, 179, 183
Fox, N. A. 136 - 137, 183
Fragaszy, D. M. 278
Franceschi, C. 276
Franchin, L. 17, 116, 131, 140, 182, 276
Francini, M. 18, 192, 227, 276
Frank, M. 136, 184
Frazer, J. 62, 207, 270
Fredegar 228
Freedman, R. A. 258, 271
Fridlund, A. J. 135
Friedman, L. M. 9, 80
Friedman, J. B. 261, 270
Friesen, W. V. 133, 135 - 136, 183 - 184
Frijda, N. H. 131 - 132, 183, 185
Frisk, H. 213, 270

Gaia 196, 199, 206
Gaimar 236
Galilei, G. 246
Galison, P. 266, 270
Gamble, C. 53, 110
Gardner, A. R. 157, 184
Gardner, B. 157, 184
Garmonsway, G. N. 235, 270
Gasser, J. 100, 107
Gaynor, D. 123, 182
Geb 198
Gellner, E. 12, 264 - 265
Gellrich, J. 244, 253, 270
Geymonat, L. 205, 273
Gianfranchi, C. 140, 182
Gildas 229 - 231
Gillispie, C. C. 174, 177, 184
Gomulicki, B. R. 100, 107
Gonçalves, L. 145, 184
Goody, J. 164, 184
Graffi, G. 148, 184
Grahn, D. 32, 106
Gransden, A. 229, 232 - 233, 236 - 237, 239, 270
Gravina, B. 51, 107, 109
Green, R. E. 51, 107 - 108
Greenacre, P. 87, 108
Greene, T. M. 66
Gregory of Tours 228
Gregory the Great (pope) 230 - 231
Griffin, A. S. 123, 182
Griffith, T. 58, 108
Grignolio, A. 17, 27, 81, 277
Groningen, B. A. van 195, 270
Grün, R. 47, 111
Guattari, F. 165, 182

Haak, W. 54, 108
Haeckel, E. 97, 99, 102
Hager, J. C. 133, 183
Halbwach, M. 227, 270
Hamblin, K. W. 260, 271
Hamilton, D. 174, 184
Hamilton, W. D. 123 - 124, 174, 184

Hans 221
Hanus, D. 127, 187
Harding, A. J. 91, 108
Hare, B. 127, 187
Harmon, R. J. 137, 138, 183
Harries, K. 108
Hart, T. H. 97 - 98, 108
Hartog, M. 97, 100, 106
Hauck, K. 240, 271
Haupt, H. 178
Hauser, M. D. 154, 156, 157, 159 - 160, 183 - 185
Hayes, C. 157, 184
Hayes, D. P. 38, 108
Hayflick, L. 35, 108
Hebden, J. C. 145, 184
Heemst, D. van 33, 108
Hegel, G. W. F. 15 - 16
Heidegger, M. 16, 72 - 73, 108, 170
Heine, B. 150
Helghesson, O. 219
Hembree, E. A. 133, 185
Henderson, J. 200, 271
Hengest 233, 239
Heracles 209
Hering 28, 81, 97, 100 - 102
Herodotus 58, 108, 205
Herrenschmidt, C. 219, 271
Hesiod 13, 14, 191, 196 - 197, 199, 205, 207
Heyne, C. 173 - 174
Higham, N. J. 232, 271
Hill, B. 87, 108
Hjorstijo, C. H. 133, 184
Hobbes, T. 175
Hoffner, H. A. 271
Hollander, L. M. 217, 271
Holt, L. E. 83, 108
Homer 197 - 198, 205
Horace 262 - 264, 266
Horsa 233, 239
Hoskin, K. 18, 117, 163, 165, 167, 171 - 172, 177 - 178, 183 - 184, 277
Hsu, H. C. 136, 183
Hudjashov, G. 49, 108
Hughes, D. H. 12
Hull, R. F. C. 16
Hume, D. 175
Hunter, P. 255, 271
Huntingford, F. A. 120, 184
Hurford, J. R. 148, 184
Hurles, M. E. 52, 108
Hutchinson, D. S. 106
Huxley, T. H. 86 - 87, 129
Huyghe, P. 223 - 225, 271
Hyppocrates 205
Hyppolytus of Rome 206

Ibn Arabi 14
Inwood, M. J. 16
Isis 198
Izard, C. E. 133, 139, 184 - 185

Jackendoff, R. 149, 159 - 161, 182, 185, 186
Jacob 244
Jardine, G. 174
Job 213
Jobling, M. A. 52, 108
John (saint) 243
Johnson, E. G. 249, 271
Johnson, T. E. 34, 108
Jones, B. 144, 176, 185
Jones, H. F. 93, 106
Jones, R. 173
Jones, W. H. S. 207, 271
Jordanes 228
Jost, J. 253, 269
Joyce, J. 262 - 264

Kail, R. 135, 185
Kako, E. T. 158, 185
Kâla 202
Kamil, A. C. 122, 181
Kanade, T. 145, 182, 187
Kansky, R. 123, 182
Kanzi (chimpanzee) 157
Kapoor, A. 9, 77, 79
Kaufmann, W. J. III 258, 271
Kern, O. 208, 271
Keynes, G. 87, 108
Keynes, S. 239, 271
Kimmelman, M. 74, 108
Kirby, S. 150, 182

Kirkwood, T. B. L. 32, 108
Kivisild, T. 41, 111
Klaeber, F. 239, 271
Kleingünter, A. 208, 271
Knight, C. 148, 150, 184
Koenig, K. L. 138, 183
Kølvraa, S. 34, 110
Korner, A. F. 138, 185
Kozlowski, J. 52, 108
Krause, Dr. 88
Krings, M. 51, 108
Kristeller, P. O. 67
Kuhn, H. 217, 263, 272
Kuhn, T. S. 259, 263, 267, 271
Kumarbi 199
Kuningas, M. 33, 108
Kurjak, A. 145, 187
Kuteva, T. 150

Lagercrantz, H. 145, 181
Laks, A. 271
Lamarck, J.-B. 28, 82, 86, 88, 90, 97-98, 102
Lamb, W. R. M. 207, 271
Lamberti, E. 276
Lambeth, S. P. 127, 186
Langford, D. J. 128, 185
Lapidge, M. 239, 271
Latham, R. E. 231, 273
Le Bourg, E. 37, 109
Lee, W. 145, 184
Leeuwenhoek, A. van 266
Lehmann, W. P. 215, 271
Lenneberg, E. H. 150, 185
Levenstadt, J. S. 128, 185
Leventhal, R. 173, 185
Leviathan 175
Levitin, D. J. 128, 185
Levy, E. K. 109
Lewis, M. 141, 185
Lewis, W. 263
Lewis-Williams, D. 64-65, 75, 109
Liberman, A. M. 158, 185
Lichtheim, M. 271
Lidenbrock, A. 221
Lidenbrock, O. 221
Lien, J. 145, 182
Lincoln, B. 219, 271
Linus 209
Liska, V. 263
Loertscher, M. 58
Lombardi, O. 140, 182
Lorenz, K. 120
Lubbock, J. 52, 96, 109
Lucian 262
Lund, N. 239, 272
Luxen, A. 144, 185
Lycurgus 208
Lyell, C. 96, 109
Lyons, J. 12, 265

Maagaard, C. 244, 269
Macaulay, V. 41, 48, 105, 109
Macfarlane, R. 92, 96, 109
MacFie, A. L. 175-176, 186
Macve, R. 167, 177-179, 184
Mægla 234
Maggia, F. 222, 271
Malleson, F. A. 221, 273
Malone, E. 249
Manes 58
Mann, J. 250, 269, 271
Manser, M. B. 123, 182
Manstead, A. S. R. 131, 185
Maquet, P. 144, 185
Marcovich, M. 271
Márka, M. 34, 106
Marks, G. A. 144, 185
Martian 230
Martin, G. M. 33, 109
Mary (virgin) 250, 252
Mascaro, J. 127, 186
Matsumura, S. 48, 107
Mauricius 233
Mayr-Harting, H. 232, 272
Mc Cartney, R. D. 138, 183
McGrew, W. C. 129, 185
McIlrath, G. M. 123, 182
Mecon 207
Meli, M. 217, 272
Melis, A. P. 127, 187
Mellars, P. 47, 49, 51, 107, 109
Mengal, P. 99

Menozzi, P. 47, 54, 106, 109
Messer-Davidow, E. 172, 184
Messinger, D. S. 135 - 137, 139 - 140, 182, 185, 276
Michelet, F. L. 231, 236 - 237, 272
Milton, J. 253 - 254
Mirmiran, M. 144, 186
Mithen, S. 64 - 65, 75, 109
Mogil, J. S. 128, 185
Moisl, H. 238, 272
Monge, G. 173, 177
Monier-Williams, M. 212, 272
Mooijaart, S. P. 33, 108
Morse, S. F. B. 179
Mosedale, D. 131
Most, G. W. 14, 196, 271 - 272
Munro, T. 66
Murray, A. T. 198, 272
Mustica, T. 145, 181
Myers, M. L. 175, 176, 186

Naassene 206
Nadar, F. 133
Natanleod 234
Neckel, G. 217, 272
Nefti 198
Nei, M. 47, 109
Nelson, J. L. 239, 272
Nelson-Goens, G. C. 136, 140, 183, 185
Nespor, M. 275
Newton, I. 259, 266
Niedermayr, W. 222 - 223
Noonan, J. P. 51, 109
Norman, R. 85, 109
Nouvel, J. 9, 75
Nut 198

Oceanus 198
Oðinn 219
O'Donnell, T. 157, 184
Oksenberg, A. 144, 185
Ólafr Tryggvasson 217
Olivieri, A. 50, 58, 109
Olshansky, S. J. 32, 36, 106, 109
Ong, W. 164, 186
O'Riain, M. J. 123, 182
Origen of Alexandria 213, 272
Orosius 230
Orsini, E. 213, 272
Osborne, H. 247, 272
Osburh 238
Osgood, C. W. 246, 272
Osiris 198
Oslac 238
Oster, H. 134, 139, 141, 186
O'Sullivan, M. 143, 183
Otis, L. 83, 90, 109 - 110
Otte, M. 52 - 53, 108, 110
Ouranos 196, 199, 206
Ovid 219

Pagani, I. 219, 272
Pala, M. 58
Palamedes 208
Paley, W. 93
Pan Gu 201
Pandora 191, 207
Papendieck, G. 145, 181
Paradis, J. G. 84, 93, 110
Parkin-Gounelas, R. 84, 101
Pascal, B. 15
Paul the Deacon 228
Pausanias Periegeta 207
Pazzini, D. 213, 272
Penglase, C. 272
Pepperberg, I. M. 122, 181
Perego, U. A. 17, 25, 41, 277
Perletti, G. 91, 110
Perls, T. T. 36, 109
Persephones 208
Péters, J. M. 144, 185
Pfeifer, E. J. 98, 110
Pherecydes of Syrius 201 - 202
Phoroneus 207
Piattelli-Palmarini, M. 150 - 151, 186
Piazza, A. 47, 54, 106, 109
Picasso, P. 263
Pilbeam, D. 12
Pinker, S. 148 - 150, 159 - 160, 185 - 186
Plassmann, A. 231, 240, 272
Plato 13, 61, 206 - 207
Pliny 230

Plummer, C. 232 - 233, 239, 272
Pokorny, J. 212, 272
Port 234 - 235
Poseidon 198
Pound, E. 262
Povinelli, D. J. 127, 186
Prajapati 201
Prechtl, H. F. R. 142, 186
Prigogine, I. 12
Pritchard, J. B. 272
Prometheus 191, 207 - 208
Ptolemy 253
Pulsiano, P. 232, 273
Purusha 201
Pyndarus 209

Quintana-Murci, L. 47, 110

Rabissoni, M. 140, 182
Raby, P. 83, 86, 110
Ramsey, C. B. 51, 107, 109
Raphael, D. D. 175 - 176, 186
Raswen, G. 219
Rattan, S. I. S. 18, 25, 29, 32 - 35, 37 - 38, 105, 109 - 110, 277
Rattray, R. F. 89, 94, 98 - 99, 110
Rawlings, G. B. 15
Rawlinson, G. 48, 108
Reed, L. I. 145, 186
Rees, M. J. 12
Reynolds, J. sir 249
Rhea 196
Ricci, P. G. 245, 272
Richards, M. 41, 55, 105, 110
Richardson, A. S. 110, 127, 186
Richardson, G. A. 67, 111
Richardson, N. 206, 272
Rilke, R. M. 262
Rizzi, L. 153, 186
Roberts, A. J. 206, 272
Robertson, D. W. 246, 272
Robinson, R. 93
Roffwarg, H. P. 144, 185
Romanes, G. 86 - 87
Romero, R. 145, 184
Rose, M. R. 32, 110
Rosenberg, E. 135, 186
Rosenblum, L. A. 141, 185
Ross, W. D. 110, 202
Rowlandson, T. 261
Roychoudhury, A. K. 47, 109
Russell, J. A. 132, 135, 186

Salda, M. 253, 269
Sanders, A. 95, 110
Sartre, J.-P. 16 - 17
Savage-Rumbaugh, E. S. 157, 186
Saxo Grammaticus 220
Scarpe, L. 123, 182
Schacter, D. L. 101, 110
Schaik, C. P. van 126, 181
Schapiro, S. J. 127, 186
Schelling, F. D. J. 61, 110
Schmandt-Besserat, D. 164, 186
Schmidt, K. L. 145, 186
Schönberg, A. 262
Schulkin, J. 30, 110
Schützeichel, R. 214, 272
Schwyzer, E. 213, 273
Scragg, D. C. 232, 273
Secrist, C. 136, 183
Segerstrale, U. 120, 186
Semsei, I. 34, 106
Serra, G. 145, 181
Seth 198
Shaffery, J. P. 144, 185
Shapiro, A. F. 136, 183
Sharpe, L. 123, 182
Shaw, G. B. 85, 110
Shehzad, Z. 128, 185
Shelton, A. 106
Sherley-Price, L. 231 - 232, 273
Shu 198
Shuttleworth, S. 84, 96, 97, 101, 111
Silk, J. B. 127, 186
Simion, F. 140, 182, 185, 276
Simpson, J. A. 259, 273
Sims-Williams, P. 235, 273
Singh, R. 34, 110
Sisam, K. 238, 273
Skilton, D. 278
Slagboom, P. E. 33, 108

Slote, S. 262 – 263, 273
Smith, A. 174 – 176, 186, 227, 273
Smith, J. A. 110, 202, 273
Smith, J. M. 12
Smith, S. B. 128, 185
Smyth, A. P. 237, 273
Snorri Sturluson 217 – 218, 221
Soffer, O. 53, 110
Sorel, A. 250
Sotocinal, S. G. 128, 185
Souriau, E. 66
Speciale, S. G. 144, 185
Spencer, H. 66, 170
Sperber, D. 155, 186
Spinozzi, P. 11, 17, 26, 85, 106, 276, 278
Spitz, R. A. 137, 186 – 187
Sroufe, L. A. 137 – 138, 187
Sterling, P. 30
Stevenson, W. H. 238, 273
Stoneking, M. 43, 48, 106
Streatfeild, R. A. 93, 97, 106
Stringer, C. B. 43, 47, 111
Studdert-Kennedy, M. 148, 184
Stuf 234
Swanton, M. J. 232, 236, 239, 273
Syntax, Dr. 261

Tabasso, A. 140, 182, 276
Tanchumah bar Abba 14
Tattersall, I. 156, 161, 187
Taylor, J. B. 84, 96, 111
Taylor, F. W. 178
Tefnut 198
Tesky, G. 254, 273
Tethys 198
Thales 203 – 204
Thangaraj, K. 48, 111
Thayer, S. 177 – 178
Thelen, E. 139, 183
Tholomeus III 209
Thomas of Aquinas 248
Thomas, J. 278
Thorne, A. G. 43, 47, 111
Þorsteinsson, J. G. 193, 257, 278
Thucydides 205
Tiamat 198, 201
Tian, Y.-I. 145, 187
Tomasello, M. 127 – 128, 150, 187
Torroni, A. 42, 46, 54 – 55, 111
Tournachon, A. 133
Treharne, E. 232, 273
Trousson, R. 85, 106, 276, 278
Trykkä, A. 222 – 223
Tcherkassof, A. 132, 184
Turner, A. 58, 273
Tyler, D. 178
Tyler-Smith, C. 52, 108
Tyrrhênus 58

Ujhelyi, M. 150, 185
Underhill, P. A. 41, 111
Urbán, L. 34, 106

Vacca, T. 140, 182
Valentinian 230, 233
Valerian 251
Vasey, P. L. 124, 187
Vaupel, J. W. 34, 108
Vé 219
Vegetti, M. 205, 273
Verne, J. 192, 221 – 224, 273
Vicky (chimpanzee) 157
Vigfusson, G. 214, 270
Vilhjálmsson, Þ. 278
Vili 219
Vinci, L. da 61
Visalberghi, E. 278
Vitale, A. 17, 115, 119, 278
Volkelt, J. 66
Vonk, J. 127, 186
Vortigern 230, 233
Vrtra 201
Vasey, P. L. 187

Waal, F. B. de 115, 126 – 129, 181, 185, 187
Wais, K. 67, 111
Walusinski, O. 145, 187
Warneken, F. 127 – 128, 187
Warner, H. 33, 111
Washoe (chimpanzee) 157
Waters, E. 111, 138, 187
Watson, E. 47, 111

Watt, J. 179
Wawn, A. 221, 273
Weinberg, S. 265 – 267, 273
Weiner, E. S. C. 259, 273
Weismann, A. 32, 100, 108
Wenskus, R. 228, 273
West, M. L. 197, 199, 202, 273
Westendorp, R. G. J. 33, 109
Whistler, G. 178
Widukind of Corvey 228
Wihtgar 234
Wilkins, J. (bishop) 169
Willey, B. 88, 98, 111
Williamson, G. S. 91, 111
Wilson, A. C. 43, 106
Wilson, D. 155, 186
Wilson, E. O. 120, 187
Winberg, J. 145, 181
Winterson, J. 244, 273
Wipped 233
Wittgenstein, L. 263 – 264, 273
Wlencing 233
Wodan 216
Woden 238
Wolf, F. 173 – 174
Wolf, K. M. 137, 187
Wolff, P. H. 137 – 139, 187
Wolpoff, M. H. 43, 111
Wood, I. 239
Woolf, V. 85, 111
Wormald, P. 232, 274
Wray, A. 150, 187

Yale, M. E. 136
Yates, F. E. 30, 111
Ymir 201, 217, 219 – 220

Zas 201 – 202
Zeher, M. 34, 106
Zeus 196 – 199, 201, 207 – 208
Zironi, A. 11, 18, 192, 211, 214, 274, 278
Zlochower, A. J. 145, 182
Zumthor, P. 235, 274
Zurvan 202
Zwaan, B. J. 33, 108